PRINCIPLES OF STATISTICS

M. G. BULMER

M.A., D.PHIL.

Fellow of Wolfson College
and Lecturer in Biomathematics, University of Oxford

DOVER PUBLICATIONS, INC., NEW YORK

Published in Canada by General Publishing Company, Ltd., 30 Lesmill Road, Don Mills, Toronto, Ontario.
Published in the United Kingdom by Constable and Company, Ltd.

This Dover edition, first published in 1979, is an unabridged and corrected republication of the second (1967) edition of the work first published by Oliver and Boyd, Edinburgh, in 1965.

International Standard Book Number: 0-486-63760-3
Library of Congress Catalog Card Number: 78-72991

Manufactured in the United States of America
Dover Publications, Inc.
180 Varick Street
New York, N.Y. 10014

CONTENTS

PREFACE TO FIRST EDITION

The aim of this book is to describe as simply as possible the fundamental principles and concepts of statistics. I hope that it may be useful both to the mathematician who requires an elementary introduction to the subject and to the scientist who wishes to understand the theory underlying the statistical methods which he uses in his own work. Most of the examples are of a biological nature since my own experience and interests lie in this field.

I am grateful to the *Biometrika* Trustees for permission to reproduce Fig. 15 and the tables at the end of the book, and to Professor E. W. Sinnott and the National Academy of Sciences for permission to reproduce Fig. 23.

I wish to thank Mr. J. F. Scott for reading and criticising the manuscript and Mrs. K. M. Earnshaw for assistance with the typing.

M. G. B.

Oxford, 1965

PREFACE TO SECOND EDITION

In this edition I have added a number of exercises and problems at the end of each chapter. The exercises are straightforward and are mostly numerical, whereas the problems are of a more theoretical nature and many of them extend and expand the exposition of the text. The exercises should not present much difficulty but the problems demand more mathematical maturity and are primarily intended for the mathematician. However, I recommend mathematicians to do the exercises first and to reserve the problems until a second reading since it is important for them to acquire an intuitive feeling for the subject before they study it in detail.

Some of the exercises and problems are taken from the Preliminary Examination in Psychology, Philosophy and Physiology and from the Certificate and Diploma in Statistics of Oxford University and I am grateful to the Clarendon Press for permission to print them.

I have also taken the opportunity to rewrite the sections on correlation in Chapters 5 and 12, and to correct a number of misprints, for whose detection I am indebted to Professor P. Armitage, Miss Susan Jennings, and a number of other readers.

M. G. B.

Oxford, 1966

Chapter 1

THE TWO CONCEPTS OF PROBABILITY

> "*But 'glory' doesn't mean 'a nice knock-down argument'," Alice objected.*
>
> "*When* I *use a word," Humpty Dumpty said, in rather a scornful tone, "it means just what I choose it to mean—neither more nor less.*"
>
> "*The question is," said Alice, "whether you* can *make words mean so many different things.*"
>
> "*The question is," said Humpty Dumpty, "which is to be master—that's all.*"
>
> Lewis Carroll: *Through the Looking-glass*

It is advisable in any subject to begin by defining the terms which are to be used. This is particularly important in probability theory, in which much confusion has been caused by failure to define the sense in which the word *probability* is being used. For there are two quite distinct concepts of probability, and unless they are carefully distinguished fruitless controversy can arise between people who are talking about different things without knowing it. These two concepts are: (1) the relative frequency with which an event occurs in the long run, and (2) the degree of belief which it is reasonable to place in a proposition on given evidence. The first of these I shall call *statistical probability* and the second *inductive probability*, following the terminology of Darrell Huff (1960). We will now discuss them in turn. Much of this discussion, particularly that of the logical distinction between the two concepts of probability, is based on the excellent account in the second chapter of Carnap (1950).

Statistical Probability

The concept of statistical probability is based on the long run stability of frequency ratios. Before giving a definition

I shall illustrate this idea by means of two examples, from a coin-tossing experiment and from data on the numbers of boys and girls born.

Coin-tossing. No one can tell which way a penny will fall; but we expect the proportions of heads and tails after a large number of spins to be nearly equal. An experiment to demonstrate this point was performed by Kerrich while he was interned in Denmark during the last war. He tossed a coin 10,000 times and obtained altogether 5067 heads; thus at

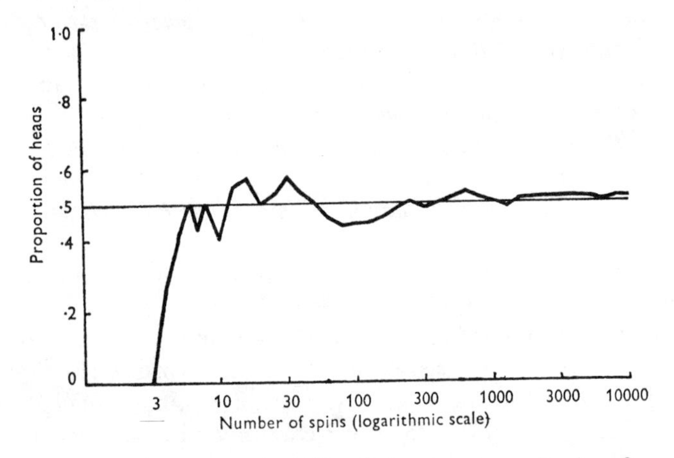

FIG. 1. The proportion of heads in a sequence of spins of a coin (Kerrich, 1946)

the end of the experiment the proportion of heads was ·5067 and that of tails ·4933. The behaviour of the proportion of heads throughout the experiment is shown in Fig. 1. It will be seen that it fluctuates widely at first but begins to settle down to a more or less stable value as the number of spins increases. It seems reasonable to suppose that the fluctuations would continue to diminish if the experiment were continued indefinitely, and that the proportion of heads would cluster more and more closely about a limiting value which would be very near, if not exactly, one-half. This hypothetical limiting value is the (statistical) probability of heads.

The sex ratio. Another familiar example of the stability of frequency ratios in the long run is provided by registers of births. It has been known since the eighteenth century that in reliable birth statistics based on sufficiently large numbers there is always a slight excess of boys; for example Laplace records that among the 215,599 births in thirty districts of France in the years 1800-1802 there were 110,312 boys and 105,287 girls; the proportions of boys and girls were thus ·512 and ·488 respectively. In a smaller number of births one would, however, expect considerable deviations from these proportions. In order to give some idea of the effect of the size

TABLE 1

The sex ratio in England in 1956
(Source: *Annual Statistical Review*)

Regions of England	Sex ratio	Rural districts of Dorset	Sex ratio
Northern	·514	Beaminster	·38
E. and W. Riding	·513	Blandford	·47
North Western	·512	Bridport	·53
North Midland	·517	Dorchester	·50
Midland	·514	Shaftesbury	·59
Eastern	·516	Sherborne	·44
London and S. Eastern	·514	Sturminster	·54
Southern	·514	Wareham and Purbeck	·53
South Western	·513	Wimborne and Cranborne	·54
Whole country	·514	All R.D.'s of Dorset	·512

of the sample on the variability of the sex ratio I have calculated in Table 1 the proportions of male births in 1956 (*a*) in the major regions of England, and (*b*) in the rural districts of Dorset. The figures for the major regions of England, which are each based on about 100,000 births, range between ·512 and ·517, while those for the rural districts of Dorset, based on about 200 births each, range between ·38 and ·59. The larger sample size is clearly the reason for the greater constancy of the former. We can imagine that if the sample were increased indefinitely, the proportion of boys would tend to a limiting value which is unlikely to differ much from ·514, the sex ratio for the whole country. This hypothetical limiting value is the (statistical) probability of a male birth.

Definition of statistical probability

A statistical probability is thus the limiting value of the relative frequency with which some event occurs. In the examples just considered only two events were possible at each trial: a penny must fall heads or tails and a baby must be either a boy or a girl. In general there will be a larger number of possible events at each trial; for example, if we throw a die there are six possible results at each throw, if we play roulette there are thirty-seven possible results (including zero) at each spin of the wheel and if we count the number of micro-organisms in a sample of pond water the answer can be any whole number. Consider then any observation or experiment which can, in principle at least, be repeated indefinitely. Each repetition will result in the occurrence of one out of an arbitrary number of possible outcomes or events, which will be symbolised by the letters A, B, C and so on. If in n repetitions of the experiment the event A has occurred $n(A)$ times, the proportion of times on which it has occurred is clearly $n(A)/n$, which will be denoted by $p(A)$. In many situations it is found that as the number of repetitions increases $p(A)$ seems to cluster more and more closely about some particular value, as does the proportion of heads in Fig. 1. In these circumstances it seems reasonable to suppose that this behaviour would continue if the experiment could be repeated indefinitely and that $p(A)$ would settle down with ever diminishing fluctuations about some stable limiting value. This hypothetical limiting value is called the (statistical) probability of A and is denoted by $P(A)$ or $Prob(A)$.

Two important features of this concept of probability must be briefly mentioned. Firstly, it is an empirical concept. Statements about statistical probability are statements about what actually happens in the real world and can only be verified by observation or experiment. If we want to know whether the statistical probability of heads is $\frac{1}{2}$ when a particular coin is thrown in a particular way, we can only find out by throwing the coin a large number of times. Considerations of the physical symmetry of the coin can of course provide good, *a priori* reason for conjecturing that this probability is about $\frac{1}{2}$, but confirmation, or disproof, of this conjecture can only come from actual experiment.

Secondly, we can never know with certainty the exact probability of an event. For no experiment can, in practice, be continued indefinitely, since either the apparatus or the experimenter will wear out sooner or later; and even if it were possible to repeat the experiment for ever, we could never reach the end of an endless sequence of relative frequencies to find out what their limiting value is.* It follows that the above definition of statistical probability cannot be interpreted in a literal, operational sense. Some authors, such as Jeffreys (1961), have concluded that the concept of statistical probability is invalid and meaningless; but the philosophical difficulties in defining probability are no greater than those encountered in trying to define precisely other fundamental scientific concepts such as time and should not prevent us from using this concept, whose meaning is intuitively clear. The reader who wishes to pursue this topic further is referred to the books of Braithwaite (1955), von Mises (1957) and Reichenbach (1949).

Inductive Probability

The second concept of probability is that of the degree of belief which it is rational to place in a hypothesis or proposition on given evidence. J. S. Mill gives a very clear definition of this concept: "We must remember," says Mill, "that the probability of an event is not a quality of the event itself, but a mere name for the degree of ground which we, or someone else, have for expecting it. . . . Every event is in itself certain, not probable: if we knew all, we should either know positively that it will happen, or positively that it will not. But its probability to us means the degree of expectation of its

* It should be noted that the concept of the limit of a mathematical series cannot be extended to an empirically given series of relative frequencies. A mathematical limit only exists because the series is determined in advance by a rule of formation whereas it is in the nature of an empirical series that it is not known in advance.

occurrence, which we are warranted in entertaining by our present evidence." *

This concept of probability should be familiar to most people; for, as Bishop Butler wrote, " To us, probability is the very guide of life." It is perhaps most clearly illustrated in the deliberations of juries. The function of the jury in a criminal prosecution is to listen to the evidence, and then to determine the probability that the prisoner committed the crime of which he is accused. If they consider the probability very high they bring in a verdict of guilty; otherwise, a verdict of not guilty. In a civil action, on the other hand, the jury, if there is one, will find for the party which they consider to have the higher probability of being correct in its assertions. The probability which it is the function of the jury to assess is clearly not a statistical probability; for each trial is unique and cannot be considered as one out of a large number of similar trials. What the jury does is to decide, after hearing the evidence, what, as reasonable men, they ought to believe and with what strength they should hold that belief. This concept will be called *inductive probability*.

The essential difference between the two concepts of probability is that statistical probability is, as we have already seen, an empirical concept while inductive probability is a logical concept. Statements about inductive probability are not statements about anything in the outside world but about the logical relationship of partial implication between a proposition or hypothesis in which we believe more or less strongly and the evidence on which that belief is based. They therefore belong to the branch of Logic called Inductive Logic which concerns partial implication, just as Deductive Logic concerns complete implication.

* *Logic*, Book 3, Chapter 18. It is interesting to note that in the first edition Mill adopted a frequency definition of probability. (" Why," he asks, " in tossing up a halfpenny, do we reckon it equally probable that we shall throw cross or pile? Because we know that in any great number of throws, cross and pile are thrown equally often; and that the more throws we make, the more nearly the equality is perfect.") In later editions this frequency approach is explicitly rejected and the definition of inductive probability quoted above is substituted for it. Mill did not realise that the two concepts of probability are equally valid and not incompatible.

The non-empirical nature of inductive probability is shown clearly by the fact that if we wish to know the degree of belief we are entitled to place in a hypothesis on the evidence to hand all we can do is to think about the problem; we are debarred from collecting fresh evidence since we should then have a new probability based on different evidence. It is important to remember that inductive probabilities depend on the evidence and change when new evidence is obtained. For example, it is now known with virtual certainty that the Piltdown skull is a fake, and consists of an ancient human skull together with the lower jaw of a chimpanzee coloured to resemble it (Weiner, 1955); but before this was proved it was quite reasonable to believe the skull to be genuine and to try to fit it into the general scheme of human evolution.

We have seen, therefore, that there are two distinct concepts of probability; we must now consider how these concepts are to be used in the theory of statistics. When probability occurs as part of a scientific or statistical hypothesis it is clearly being used in its empirical, frequency sense; for such hypotheses are by nature empirically testable. For example, when we ask, " Is this coin unbiased? ", that is to say, " Is the probability of heads $\frac{1}{2}$? ", we are using the statistical sense of probability; for the question can only be answered by spinning the coin a large number of times. When, however, we come to statistical inference and ask questions like, " Given that this coin has shown 53 heads and 47 tails in 100 throws, is this good evidence for believing it to be unbiased? ", we are using the idea of rational degree of belief, and the answer might be, " There is a high probability (in the inductive sense) that the probability of heads (in the statistical sense) is $\frac{1}{2}$ ". Again, it is quite clear that the probabilities which occur in Mendelian genetics are statistical probabilities; but if we were to ask, " How strongly do Mendel's results confirm his theory? " we should be seeking information about the rational degree of belief to be placed in that theory.

It would seem, therefore, that both concepts of probability have their part to play in the theory of statistics, statistical probability when we are formulating statistical hypotheses and inductive probability when we make statistical inferences about the adequacy of such hypotheses to fit the facts. However, orthodox statistical opinion declines to use inductive probabilities in the latter context on the ground that they are not quantifiable and so cannot be employed in a mathematical argument, and the usual methods of statistical inference such as significance tests and confidence intervals are based entirely on the concept of statistical probability. We must, therefore, now consider why it is difficult to express inductive probabilities numerically.

The Measurement of Inductive Probabilities

Most attempts to construct a numerical scale of inductive probabilities, with 0 standing for impossibility and 1 for logical certainty, start from the Principle of Indifference, which states that two events are equally probable if we have no reason to suppose that one of them will happen rather than the other. This principle was formulated as follows by Laplace in 1814:

"The theory of chances consists in reducing all events of the same kind to a certain number of cases equally possible, that is, such that we are *equally undecided* as to their existence; and in determining the number of these cases which are favourable to the event of which the probability is sought. The ratio of that number to the number of all the possible cases is the measure of the probability; which is thus a fraction, having for its numerator the number of cases favourable to the event, and for its denominator the number of all the cases which are possible."

The Principle of Indifference seems at first sight very plausible. If we know that one of two events must occur but have no reason to suppose that it will be one of them rather than the other, what else can we do but attribute a probability of $\frac{1}{2}$ to each of them? For it would clearly be illogical to attribute a higher probability to one of them than to the other. Once we admit that all inductive probabilities can be measured we are driven to this view. There are, however, two strong objections to it as a basis for constructing a scale of probabilities.

In the first place the Principle of Indifference can only be applied when the alternatives can be split up into a number of equally possible cases; this is a severe limitation on its usefulness. Suppose for instance that we are considering the throw of a die. If we are given a die about which nothing is known except that it appears to be symmetrical, we can argue that each of the faces has an equal probability of being thrown of $\frac{1}{6}$. But suppose that we are given an asymmetrical die; we should probably be able to deduce that some of its faces were more likely to occur than others, though we could not say by how much. Or suppose we are given an apparently symmetrical die and find that in 100 throws six has occurred 22 times. This has clearly increased our belief that six will occur on the next throw; but by how much? Or again, consider a horse race. We would certainly not consider that the horses were all equally likely to win and it is difficult to see how the events could be reduced to a number of equally possible cases. The Principle of Indifference is thus of very limited applicability.

The second objection to this principle is that even when it can be applied it leads to inconsistencies and paradoxes. Suppose that we are given a glass containing a mixture of wine and water, and that all we know about the mixture is that the proportion of water to wine is somewhere between 1:1 and 2:1; then we can argue that the proportion of water to wine is as likely to lie between 1 and $1\frac{1}{2}$ as it is to lie between $1\frac{1}{2}$ and 2. Consider now the ratio of wine to water. This quantity must lie between $\frac{1}{2}$ and 1, and we can use the same argument to show that it is equally likely to lie between $\frac{1}{2}$ and $\frac{3}{4}$ as it is to lie between $\frac{3}{4}$ and 1. But this means that the water to wine ratio is equally likely to lie between 1 and $1\frac{1}{3}$ as it is to lie between $1\frac{1}{3}$ and 2, which is clearly inconsistent with the previous calculation.

This paradox, known under the general name of Bertrand's paradox, depends on the fact that we were considering a quantity that could vary continuously between certain limits. However, similar inconsistencies arise even in the case of discrete quantities. Consider, for example, the following paradox due to d'Alembert. Suppose we toss a coin twice; what is the probability of obtaining two heads? According to the orthodox analysis there are four equally possible cases: HH, HT, TH and TT; the probability of two heads is therefore $\frac{1}{4}$. But d'Alembert argued that HT and TH both led to the same result and should therefore only be counted once; hence, he said, the probability of two heads is not $\frac{1}{4}$ but $\frac{1}{3}$. Now it can be shown that if we adopt a frequency approach, the (statistical) probability of throwing two heads in succession is $\frac{1}{4}$ if the coin is unbiased. But what objection can be advanced against d'Alembert's analysis if we are arguing entirely from the Principle of Indifference? *

The force of these objections to the Principle of Indifference is now widely recognised, and some authors have abandoned it in favour of another method of measuring inductive probabilities which was first suggested by Ramsey (1931). Suppose that I want to find out how strongly you believe that a certain horse will win the Derby. I can do this, it is suggested, by offering you a series of bets and seeing which of them you accept. For example, if the horse is an outsider, it is likely that you would reject an offer of 10 to 1, but you might well accept an offer of 100 to 1; somewhere between these two limits there must be a line dividing bets which you would accept from bets which you would not accept. If this marginal bet is 50 to 1 then your degree of belief in the horse's chances of winning the Derby is taken to be 1/51.

This method can be elaborated so as to avoid difficulties arising from the marginal utility of money (a shilling is worth more to a beggar than to a millionaire) and from the fact that betting may in itself be pleasant or unpleasant. There are, however, two fundamental objections to it. The first is that, even if you can be forced to say that you would accept a bet at odds of 50 to 1 but that you would not accept one at odds of 49 to 1, it still seems unnatural to try to define the position of the dividing line precisely. But the more serious objection is that this method leads not to an objective scale measuring

* Objection can also be raised against the orthodox analysis on the grounds that the four cases are not equally possible. For if the result of the first throw is heads, this provides some information about the coin and should raise, even though slightly, our expectation that the second throw will be heads rather than tails.

how strongly you ought to believe in the horse's chances of success but to a subjective, psychological scale measuring your actual degree of belief. It is possible, and indeed probable, that two people on the same evidence will have quite different betting propensities. Such a psychological scale varying from one person to another is of little use in scientific discussion in which the scientist must persuade others to believe what he himself believes.

It has been reluctantly concluded by most statisticians that inductive probability cannot in general be measured and, therefore, cannot be used in the mathematical theory of statistics. This conclusion is not, perhaps, very surprising since there seems no reason why rational degrees of belief should be measurable any more than, say, degrees of beauty. Some paintings are very beautiful, some are quite beautiful and some are ugly; but it would be absurd to try to construct a numerical scale of beauty on which the *Mona Lisa* had a beauty-value of ·96! Similarly some propositions are highly probable, some are quite probable and some are improbable; but it does not seem possible to construct a numerical scale of such (inductive) probabilities. A full and most readable account of the problem of measuring such probabilities will be found in Keynes (1921).

CHAPTER 2

THE TWO LAWS OF PROBABILITY

We saw in the last chapter that there are two concepts of probability but that only the first of them, statistical probability, is capable of being expressed quantitatively. The rest of this book is concerned with the mathematical theory of statistics and consequently with statistical rather than inductive probability; probability will, therefore, be understood to mean statistical probability unless the contrary is stated.

In this chapter, we shall consider the fundamental mathematical properties of probability, which can be derived on the frequency interpretation from the consideration that what holds for a relative frequency must also hold in the limit for a probability. Thus a relative frequency must necessarily be a number between 0 and 1; hence so must a probability. Furthermore an event which cannot occur will have a probability of 0 while an event which must occur will have a probability of 1; for example, the probability that a baby will be either a boy or a girl is 1 since one or other of these events must happen. We shall now consider in turn the two main laws on which the theory of probability is based, the laws of addition and multiplication.

THE LAW OF ADDITION

The law of addition of probabilities states that if A and B are mutually exclusive events, that is if they cannot both occur together, then the probability that either A or B will occur is equal to the sum of their separate probabilities: in symbols,

$$P(A \text{ or } B) = P(A)+P(B).$$

This follows from the fact that, if A and B are mutually exclusive, the number of times on which either A or B has occurred is the number of times on which A has occurred plus the number of times on which B has occurred; the same must therefore be true of the corresponding proportions and so, ás the number of observations increases, of their limiting values or probabilities. This law can be extended to any number of events, provided they are all mutually exclusive.

For example, Table 2*a* shows the numbers of births in England and Wales in 1956 classified by (*a*) sex and (*b*) whether liveborn or stillborn; the corresponding relative

TABLE 2*a*

Numbers of births in England and Wales in 1956 by sex and whether live- or stillborn. (Source: *Annual Statistical Review*)

	Liveborn	Stillborn	Total
Male	359,881 (*A*)	8,609 (*B*)	368,490
Female	340,454 (*C*)	7,796 (*D*)	348,250
Total	700,335	16,405	716,740

TABLE 2*b*

Proportion of births in England and Wales in 1956 by sex and whether live- or stillborn. (Source: *Annual Statistical Review*)

	Liveborn	Stillborn	Total
Male	·5021	·0120	·5141
Female	·4750	·0109	·4859
Total	·9771	·0229	1·0000

frequencies are given in Table 2*b*. The total number of births is large enough for these relative frequencies to be treated for all practical purposes as probabilities. There are four possible events in this double classification, male livebirth, male still-birth, female livebirth and female stillbirth, which will be represented by the letters A, B, C and D; the compound events 'Male birth' and 'Stillbirth' will be represented by the letters M and S. Now a male birth occurs whenever either a male livebirth or a male stillbirth occurs, and so the proportion of male births, regardless of whether they are live- or stillborn, is equal to the sum of the proportions of these two types of birth; that is to say,

$$p(M) = p(A \text{ or } B) = p(A) + p(B) = \cdot 5021 + \cdot 0120 = \cdot 5141$$

where $p(A)$ means the proportion of times on which the event A has occurred. Similarly, a stillbirth occurs whenever either a male stillbirth or a female stillbirth occurs and so the proportion of stillbirths, regardless of sex, is equal to the sum of the proportions of these two events:

$$p(S) = p(B \text{ or } D) = p(B) + p(D) = \cdot 0120 + \cdot 0109 = \cdot 0229.$$

It is important to remember that the law of addition only holds when the events are mutually exclusive. Thus the proportion of times on which either a male birth or a stillbirth occurred is *not* $\cdot 5141 + \cdot 0229 = \cdot 5370$. To calculate this proportion correctly we note that a male birth or a stillbirth occurs whenever either a male livebirth or a male stillbirth or a female stillbirth occurs, and then add together the proportions of these three events:

$$\begin{aligned} p(M \text{ or } S) &= p(A \text{ or } B \text{ or } D) \\ &= p(A) + p(B) + p(D) = \cdot 5021 + \cdot 0120 + \cdot 0109 = \cdot 5250. \end{aligned}$$

What went wrong in the first calculation is that the events 'Male birth' and 'Stillbirth' are not mutually exclusive since they both occur when a boy is stillborn. In adding their proportions we have, therefore, counted the male stillbirths twice instead of once; the correct answer can be obtained by subtracting the relative frequency of this joint event ($\cdot 5370 - \cdot 0120 = \cdot 5250$).

The general rule, which can be demonstrated by a similar argument, is that if A and B are any two events, not necessarily mutually exclusive, the probability that A or B will occur is the sum of their separate probabilities minus the probability that they both occur: in symbols,

$$P(A \text{ or } B) = P(A) + P(B) - P(A \text{ and } B).$$

This rule may perhaps be made plainer by the following diagram in which the points represent the possible outcomes of some experiment each of which has a certain probability attached to it. The probability that the event A will occur is the sum of the probabilities of the points in A, and likewise for B; similarly the probability that either A or B will occur is the sum of the probabilities of points which lie in either A or B. Now if we add the probabilities of A and B we shall have counted the probabilities of all those points which are common to both A and B twice, and so

$$P(A)+P(B) = P(A \text{ or } B)+P(A \text{ and } B)$$

from which the general law of addition follows. If A and B are exclusive they will have no points in common; in this case

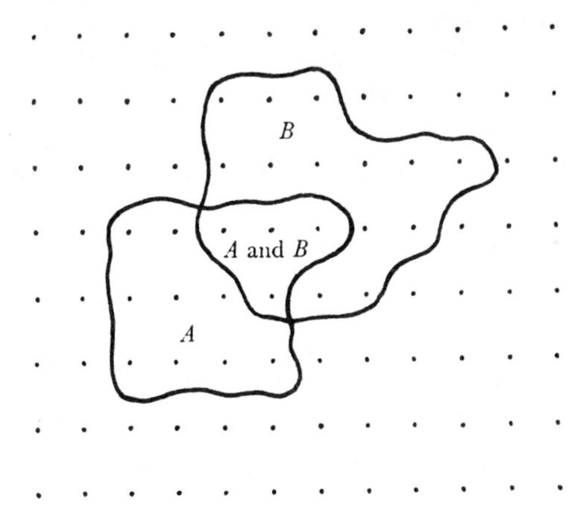

FIG. 2. Diagram to illustrate the general law of addition of probabilities

$P(A \text{ and } B) = 0$ and we again have the original form of the law. The general form of the law of addition can be extended to more than two non-mutually-exclusive events (see Problem 2.1) but the formulae become rather complicated and it is usually easier to solve problems of this sort by decomposing the events into mutually exclusive components.

THE LAW OF MULTIPLICATION

The law of multiplication of probabilities states that if A and B are two events, then the probability that both A and B

will occur is equal to the probability that A will occur multiplied by the *conditional* probability that B will occur given that A has occurred, or in symbols

$$P(A \text{ and } B) = P(A) \times P(B \mid A).$$

This law introduces the new and important concept of conditional probability which must now be considered.

The conditional probability of an event B given another event A, written $P(B \mid A)$, is the probability that B will occur if we consider only those occasions on which A also occurs; it is thus the limiting value of the proportion $n(A \text{ and } B)/n(A)$, where $n(A \text{ and } B)$ is the number of times on which both A and B have occurred and $n(A)$ is the number of times on which A has occurred. In Table 2*a* on p. 13, for example, if M represents a male birth and S a stillbirth, $n(M \text{ and } S)/n(M) = 8609/368{,}490 = {\cdot}0234$; this figure is the proportion of males who are stillborn and will, as the sample size increases, tend to a limiting value which is the probability of stillbirth in males. The corresponding proportion of stillbirths among females is $7796/348{,}250 = {\cdot}0224$.

These figures should be contrasted with the overall, or unconditional, proportion of stillbirths, which is ·0229. It will be observed that the conditional proportion of stillbirths among boys is slightly higher than, and the proportion among girls slightly lower than, the overall proportion. The difference is small, but a test of significance indicates that it represents a real difference in the corresponding probabilities (see Chapter 9, pp. 145 and 161); it can be concluded that sex and stillbirth are *statistically dependent*, that is to say that the sex of an infant has an effect, albeit a small effect, on its chance of being stillborn. This conclusion is confirmed from figures in other years and in other countries; it is in fact found that death rates at all ages are higher in men than women.

When the conditional probability of B given A is equal to the unconditional, or absolute, probability of B the two events are said to be *statistically independent*. This means that the occurrence of A does not alter the probability that B will occur. We might imagine, for example, that if we threw two dice simultaneously they would be independent of each other; for it is difficult to see how the result on one die could be affected by what happened on the other unless they were physically connected in some way. This supposition is confirmed by experience, as we shall see in the next section.

We now return to the law of multiplication. If in n repetitions of an experiment the event A has occurred $n(A)$ times and the double event A and B has occurred $n(A \text{ and } B)$ times, then it is an algebraical identity that

$$\frac{n(A \text{ and } B)}{n} = \frac{n(A)}{n} \times \frac{n(A \text{ and } B)}{n(A)}.$$

Now the left hand side of this identity is the proportion of times on which both A and B have occurred, the first term on the right hand side is the proportion of times on which A has occurred and the second term on the right hand side is the proportion of times on which B has occurred given that A has occurred. For example, the proportion of male stillbirths in Table 2*b* on p. 13 is ·0120, the proportion of male births is ·5141 and the proportion of stillbirths among the males is ·0234; it is easy to verify that ·0120 = ·5141 × ·0234. As the number of observations increases these proportions tend to the corresponding probabilities, and so

$$P(A \text{ and } B) = P(A) \times P(B \mid A).$$

If A and B are statistically independent, $P(B \mid A) = P(B)$ and the law of multiplication takes the simple form

$$P(A \text{ and } B) = P(A) \times P(B).$$

For example, if we throw two unbiased dice, that is to say two dice whose six faces have the same probability of occurring, the probability of throwing two sixes is $\frac{1}{6} \times \frac{1}{6} = \frac{1}{36}$. It is, however, important to remember that this is a special case of the law of multiplication which is only true when the events are independent.

We shall now consider some applications of these two laws of probability, first in games of chance based on the throwing of dice and then in the Mendelian theory of heredity.

Probability and Dice

The theory of probability arose in the seventeenth century from interest in games of chance. Many of these games were played with two dice, and one of the first problems which arose was to determine the frequencies with which the eleven possible scores between 2 and 12 should occur when two dice are thrown simultaneously.

Before answering this question we shall consider some actual data. Table 3 shows the results of 20,000 throws

Table 3

The results of 20,000 throws with two dice (data from Czuber, 1903)

		White die							
		1	2	3	4	5	6	Total	Proportion
Red die	1	547	587	500	462	621	690	3407	·170
	2	609	655	497	535	651	684	3631	·182
	3	514	540	468	438	587	629	3176	·159
	4	462	507	414	413	509	611	2916	·146
	5	551	562	499	506	658	672	3448	·172
	6	563	598	519	487	609	646	3422	·171
Total		3246	3449	2897	2841	3635	3932	20,000	1·000
Proportion		·162	·172	·145	·142	·182	·197	1·000	

with two dice, one white and one red, made by the Swiss astronomer Wolf in 1850. It will be seen from the marginal totals that both these dice were considerably biased. They both show a deficiency of threes and fours, and the white die also has an excess of fives and sixes at the expense of ones and twos. In seeking an explanation of this bias we must recall that dice are so constructed that opposite faces add up to 7, i.e. 1 and 6, 2 and 5, 3 and 4, are opposite each other. Thus a deficiency of threes and fours and an excess of the other numbers could be produced by a die which was not a perfect cube but was elongated in the three-four axis; the construction of false dice was well understood as far back as the sixteenth century, and such dice were called barred quater-treys (other varieties were fullams, loaded with lead, and high- and low-men, which were incorrectly numbered). The excess of 6 over 1, and to a lesser extent of 5 over 2, in the white die could be caused by the lightness of the 5 and 6 faces due to the hollowing out of the pips in the die; but it would be rash to be too certain about the appearance of dice which have long since perished. These figures do show, however, that one cannot assume that real dice are completely unbiased, although most data show considerably less bias than do Wolf's.

Despite their bias, however, the two dice are statistically independent of each other. For example, the proportion of times that the red die was 1 given that the white die was 2 is 587/3449 = ·170 which is the same as the overall proportion of times that the red die was 1. Similarly, the proportion of times that the white die was 4 given that the red die was 6 is 487/3422 = ·142 which is the same as the overall proportion of times on which the white die was 4. The reader should make some more calculations of this kind to demonstrate that the conditional proportions are in all cases nearly the same as the corresponding marginal proportions (see Exercises 2.1 and 9.9).

We now return to the problem of determining the frequencies with which the eleven possible scores between 2 and 12 should occur when two dice are thrown simultaneously. We shall suppose that the dice are unbiased and independent. In practice, as we have just seen, the assumption of independence is likely to be exactly true but the assumption of lack of bias can only be regarded as an approximation to the truth. On these assumptions the probability of obtaining a particular result, such as that the first die is 3 and the second 5, is, by the law of multiplication, 1/36. To calculate the chance of throwing a particular sum such as 8 we therefore simply count the number of ways in which that sum can be thrown and divide by 36 (by the law of addition); for example, 8 can be thrown in five ways (2—6, 3—5, 4—4, 5—3 and 6—2), and the probability of its occurrence is therefore 5/36 = ·139. The probability of obtaining other sums is calculated in exactly the same way and is set out in Table 4 together with the observed proportions calculated from Wolf's data in Table 3. It will be seen that there is quite good agreement despite the bias of Wolf's dice.

TABLE 4

The probability of throwing different sums with two dice

Sum	No. of ways of throwing	Probability	Proportion (Wolf's data)
2	1	1/36 = ·028	·027
3	2	2/36 = ·056	·060
4	3	3/36 = ·083	·083
5	4	4/36 = ·111	·098
6	5	5/36 = ·139	·134
7	6	6/36 = ·167	·166
8	5	5/36 = ·139	·139
9	4	4/36 = ·111	·108
10	3	3/36 = ·083	·088
11	2	2/36 = ·056	·064
12	1	1/36 = ·028	·032
Total	36	36/36 = 1·000	1·000

It should be noted that throws such as 4—6 and 6—4 must be counted separately since they represent different events which can be distinguished if, for example, the dice are coloured differently. The probability of throwing a four and a six with two dice, without specifying which die is the four, is therefore 2/36 since it can happen in two distinct ways; on the other hand the probability of throwing two fives is 1/36 since it can only happen in one way. As one of Damon Runyon's characters says: "Two fives is the hard way to make a ten with the dice." This point caused some difficulty in the early history of probability, and there is an interesting passage about it in the chapter on hazard in *The Compleat Gamester* by Charles Cotton (minor poet and part author of *The Compleat Angler*), which was first published in 1674. Cotton writes:

"Now six and eight one would think should admit of no difference in advantage with seven, but if you will rightly consider the case, and be so vain to make trial thereof, you will find a great advantage in seven over six and eight. How can that be you will say, hath not six, seven and eight equal chances? For example, in six, quater deuce and two treys; in eight, six deuce, cinque trey, and two quaters, and hath not seven three as aforesaid? It is confest; but pray consider the disadvantage in the doublets, two treys and two quaters, and you will find that six deuce is sooner thrown than two quaters, and so consequently, cinque Ace or quater deuce sooner than two treys: I saw an old rook once take up a young fellow in a tavern, upon this very score: the bargain was made that the rook should have seven always and the young gentleman six, and throw continually; agreed to play they went, the rook got the first day ten pound, the next day the like sum; and so for six days together losing in all threescore pounds; notwithstanding the gentleman, I am confident, had square dice, and threw them always himself."

We will now use the information in Table 4 to calculate the chance of winning at craps. Craps is a simplified version of the old game of hazard, about which Charles Cotton was writing, and which was probably introduced into Europe from the East during the Crusades; one derivation of the word is from the Arabic for a die, ' Al zhar '. Craps is played between two players, one of whom, the thrower, throws two dice. If the first throw is 7 or 11 (a natural) the thrower wins immediately; if it is 2, 3 or 12 (craps, which was called ' crabs ' in hazard), he loses immediately; if it is any other number he goes on throwing until either the same number or 7 occurs. If 7 occurs before the number he first threw he loses; otherwise, he wins.

The thrower's probability of winning obviously depends on the first throw. Thus if the first throw is 2 or 3, he loses immediately; if the first throw is 4, the probability that he will win, that is to say that he will throw 4 again before he throws 7, is 1/3 since we find from Table 4 that sevens occur twice as often as fours; similarly if the first throw is 5 the probability that he will win is 4/10; and so on. To calculate the overall probability of winning we must multiply these conditional probabilities by the probability of making the first throw in question and add them all up. The answer is

$$\frac{1}{36}\times 0+\frac{2}{36}\times 0+\frac{3}{36}\times\frac{1}{3}+\frac{4}{36}\times\frac{4}{10}+\ldots+\frac{1}{36}\times 0 = \cdot 4930.$$

The thrower is thus at a slight disadvantage. This is the reason why the player and not the bank throws the dice when craps is played in casinos. An empirical investigation has shown that in 9900 games of craps the thrower won 4871 times and lost 5029 times; the proportion of wins was thus ·4920, in excellent agreement with the theoretical calculation (Brown, 1919).

Mendel's Laws of Heredity

As a biological illustration of the laws of probability we shall consider the classical experiments of Gregor Mendel (1822-1884) on the genetics of the edible pea. In these experiments, which were carried out in the gardens of the monastery in Brünn (Brno) of which he was a member and later Abbot, Mendel began by crossing two pure lines which differed in a single contrasting character, such as a variety with purple and one with white flowers or a tall with a dwarf variety. He knew that these plants belonged to pure, inbred lines because the pea is normally self-fertilising. Mendel considered seven such characters altogether, and found that in every case the resulting hybrids resembled one of the parents; for example, all the hybrids from the cross of purple-flowered with white-flowered peas had purple flowers, and all the hybrids from the tall×dwarf cross were tall. Mendel called the

Table 5

Mendel's data on the plants bred from the hybrids. The dominant character is listed first in each case. (Source: Bateson, 1909)

Character	No. of dominants	No. of recessives	Proportion of dominants
Round *v.* wrinkled (seeds)	5,474	1,850	·747
Yellow *v.* green (seeds)	6,022	2,001	·751
Purple *v.* white (flowers)	705	224	·759
Smooth *v.* constricted (pods)	882	299	·747
Axial *v.* terminal (flowers)	651	207	·759
Green *v.* yellow (unripe pods)	428	152	·738
Tall *v.* dwarf (stem)	787	277	·740
Total	14,949	5,010	·749

character which appeared in the hybrids the *dominant* character and the one which apparently disappeared the *recessive* character. However, when these hybrids were allowed to self-fertilise the recessive character reappeared; in fact $\frac{3}{4}$ of the resulting plants had the dominant and $\frac{1}{4}$ the recessive character. His actual results are shown in Table 5.

It should be noted that this Mendelian proportion of $\frac{3}{4}$ is a (statistical) probability. Mendel himself sums this up by saying:

"These [first] two experiments are important for the determination of the average ratios, because with a smaller number of experimental plants they show that very considerable fluctuations may occur. . . .

"The true ratios of the numbers can only be ascertained by an average deduced from the sum of as many single values as possible; the greater the number, the more are merely chance effects eliminated."

Mendel explained his results by supposing that each of the seven contrasting characters was controlled by a pair of hereditary units or *genes* and that any plant contained two of these genes for each character, one derived from each of its parents. Let us consider the inheritance of flower colour as an example. If we denote the gene for purple flowers by P and that for white flowers by p, then a particular plant can have one of the three possible combinations or *genotypes*: PP, Pp or pp. When two pure lines were crossed the mating was of the type $PP \times pp$, and all the resulting plants had the genotype Pp since they received a P gene from the first and a p gene from the second parent. In order to explain why all these plants had purple flowers like the PP parent we must suppose that plants with either of the two genotypes PP or Pp have purple flowers; only pp plants have white flowers. This is quite easy to understand if we suppose that the P gene acts by catalysing the formation of a purple pigment; its effect will be the same whether it is present in single or in double dose.

Let us now see what happens when the Pp hybrids are allowed to self-fertilise. The mating is of the type $Pp \times Pp$.

In this case pollen cells containing the P and p gene will be produced in equal quantities, and likewise for the egg cells. Hence, if pollen and egg cells unite at random, that is to say independently of the gene which they contain, the following unions will occur with the same probability of $\frac{1}{2} \times \frac{1}{2} = \frac{1}{4}$:

Pollen		Egg cell		Offspring
P	$\times$	P	$=$	PP
P	$\times$	p	$=$	Pp
p	$\times$	P	$=$	Pp
p	$\times$	p	$=$	pp

Thus the genotypes PP, Pp and pp will occur in the offspring with probabilities $\frac{1}{4}$, $\frac{1}{2}$ and $\frac{1}{4}$, and so three-quarters of them will on the average have purple and one-quarter white flowers.

Mendel next proceeded to see what happened when plants differing in two contrasting characters were crossed. He therefore crossed a pure line having round, yellow seeds with a pure line having wrinkled, green seeds. All the resulting hybrids had round, yellow seeds, confirming his previous result that round seeds were dominant to wrinkled seeds and yellow seeds to green. When these hybrids in their turn were self-fertilised the results shown in Table 6 were obtained. The

TABLE 6

Mendel's data on the joint segregation of seed colour and shape. (Source: Bateson, 1909)

	Yellow	Green	Total
Round	315	108	423
Wrinkled	101	32	133
Total	416	140	556

proportion of round seeds is $423/556 = \cdot 761$, and the proportion of yellow seeds $416/556 = \cdot 748$, both near enough the theoretical value of $\cdot 75$. Furthermore we find that the proportion of round seed among the yellow seeds is $315/416 = \cdot 757$, and the proportion of round seed among the green seeds is 108/140 = ·771, which are within errors of sampling the same as the overall proportions. We conclude that the two characters behave independently, that is to say that the colour of the seed does not affect its chance of being round or wrinkled, and *vice versa*; this is known as Mendel's law of independent assortment.

Mendel did similar experiments, on a smaller scale, on several other characters; in each case he found independent assortment. However, when genetic work was resumed in 1900, it soon became clear that there were exceptions to this rule. The data in Table 7 come from a similar experiment

TABLE 7

The joint segregation of flower colour and pollen shape in the sweet pea (Bateson, 1909)

	Purple-flowered	Red-flowered	Total
Long pollen	1528	117	1645
Round pollen	106	381	487
Total	1634	498	2132

on two factors (1) purple *v*. red flower, and (2) long *v*. round pollen, in the sweet pea. Both factors give good 3:1 ratios when considered separately, but they are not independent; the proportion with long pollen is considerably larger among the purple plants than among the red ones. The reason for the dependence of the two characters was not understood until it was discovered that the genes are carried on rod-shaped bodies called chromosomes in the cell nucleus. If the genes for two different characters are carried on different chromosomes they will assort independently; if, on the other hand, they are carried in different positions on the same chromosome they will tend to be linked together, as are flower colour and pollen shape in the sweet pea. The investigation of linkage and the consequent construction of chromosome maps form an important part of modern genetics; a fuller account will be found in any textbook on genetics.

Exercises

2.1. In Table 3 on p. 18 find the conditional proportion of (*a*) 3 on red die given 6 on white die, (*b*) 5 on white die given 4 on red die, and compare them with the corresponding marginal proportions.

2.2. Find the probability that the sum of the numbers on two unbiased dice will be even (*a*) by considering the probabilities that the individual dice will show an even number, (*b*) by considering the probabilities in Table 4 on p. 20.

2.3. Find the probabilities of throwing a sum of 3, 4, ..., 18 with three unbiased dice. (This problem was considered by Galileo.)

2.4. In 1654 the Chevalier de Méré, a well-known gambler and an amateur mathematician, put a problem to Pascal which gave rise to a famous correspondence between the latter and Fermat. De Méré's problem was that he had found by calculation that the probability of throwing a six in 4 throws of a die is slightly greater than $\frac{1}{2}$, while the probability of throwing double sixes in 24 throws of two dice is slightly less than $\frac{1}{2}$; it seemed self-evident to de Méré that these probabilities should be the same and he called the result " a great scandal which made him say haughtily that the theorems were not consistent and that the arithmetic was demented " (Smith, 1929). Repeat de Méré's calculations. (It is stated in many books that de Méré had noticed this very slight inequality in the chances in actual play. There is no basis for this improbable story.)

2.5. Three men meet by chance. What are the probabilities that (*a*) none of them, (*b*) two of them, (*c*) all of them, have the same birthday? [P. P. P. Hilary, 1965].

2.6. In a certain survey of the work of chemical research workers, it was found, on the basis of extensive data, that on average each man required no fume cupboard for 60 per cent of his time, one cupboard for 30 per cent and two cupboards for 10 per cent; three or more were never required. If a group of four chemists worked independently of one another, how many fume cupboards should be available in order to provide adequate facilities for at least 95 per cent of the time? [Certificate, 1959].

2.7. A yarborough at whist or bridge is a hand of 13 cards containing no card higher than 9. It is said to be so called from an Earl of Yarborough who used to bet 1000 to 1 against its occurrence. Did he have a good bet?

Problems

2.1. For an arbitrary number of events, $E_1, E_2, \ldots, E_n$, the general law of addition states that

$$P(E_1 \text{ or } E_2 \text{ or } \ldots \text{ or } E_n) = \sum_i P(E_i) - \sum_{i<j} P(E_i \& E_j) + \sum_{i<j<k} P(E_i \& E_j \& E_k) - \ldots (-)^{n-1} P(E_1 \& E_2 \& \ldots \& E_n).$$

Prove this formula by induction. (The " or " is inclusive not exclusive.)

2.2. $E_1, E_2, \ldots, E_n$ are events; P is the probability that at least one of them occurs. Prove that

$$\sum_i P(E_i) - \sum_{i<j} P(E_i \ \& \ E_j) \leqslant P \leqslant \sum_i P(E_i).$$

An unbiased roulette wheel has 37 cups. What is the approximate probability that in 400 spins of the wheel there will be at least one cup which the ball never occupies? [Diploma, 1962]

2.3. An arbitrary number of events are said to be *pairwise independent* if

$$P(E_i \& E_j) = P(E_i) \ . \ P(E_j) \text{ for all } i \neq j.$$

They are said to be *mutually independent* if, in addition,

$$P(E_i \& E_j \& E_k) = P(E_i) \ . \ P(E_j) \ . \ P(E_k) \text{ for all } i \neq j \neq k$$

$$\vdots$$

$$P(E_1 \& E_2 \ldots \& E_n) = P(E_1) \ . \ P(E_2) \ldots P(E_n).$$

Show that pairwise independence does not imply mutual independence by supposing that two unbiased dice are thrown and considering the three events, odd number on first die, odd number on second die, odd sum on two dice.

2.4. Suppose that a genetic character is determined by a single pair of genes A and B so that there are three genotypes AA, AB and BB whose frequencies are x, y and z. Write down the expected frequencies of the different mating types, AA×AA, AA×AB, and so on, under random mating and hence show that the expected frequencies of these genotypes in the next generation under random mating are p^2, $2pq$ and q^2, where $p = x + \frac{1}{2}y$ and $q = \frac{1}{2}y + z$ are the frequencies of the A and B genes in the population. Thus the above equilibrium frequencies are attained after one generation of random mating. This law, which is fundamental in population genetics, was discovered independently by the English mathematician Hardy and the German physician Weinberg in 1908.

If the three genotypes are all distinguishable find the probabilities that (*a*) a pair of brothers, (*b*) a pair of unrelated individuals in a randomly mating population will appear the same when $p = q = \frac{1}{2}$.

2.5. The results of Table 7 on p. 25 can be explained on the assumption that the genes for flower colour and pollen shape are on the same chromosome but that there is a probability π that one of the genes will be exchanged for the corresponding gene on the other chromosome. If we denote the genes for purple or red flowers by P and p, and the genes for long and round pollen by L and l, then the hybrids from the cross considered will all be of the genotype PL/pl, the notation indicating that the P and L genes are on one chrosomome and the p and l genes on the other. When these hybrids are allowed to self-fertilise, there is a chance π that the L and l genes will interchange in one parent, giving Pl/pL; there are therefore really three mating types, $PL/pl \times PL/pl$, $Pl/pL \times PL/pl$ and $Pl/pL \times Pl/pL$, which occur with probabilities $(1-\pi)^2$, $2\pi(1-\pi)$ and π^2 respectively. Find the probabilities of the four possible phenotypes resulting from the experiment in terms of $\theta = (1-\pi)^2$.

2.6. Five players enter a competition in which each plays a game against each of the others. The two players in each game have an equal chance of winning it; a game must end in a win for one of the players who then scores a point. The competition is won by the player or players scoring most points. What is the probability that a particular player will (*a*) win the competition outright; (*b*) share in winning with other players? [Certificate, 1959]

2.7. In a game of bridge the declarer finds that he and his partner hold 9 spades between them when dummy is laid down. What are the chances that the remaining spades are distributed (*a*) 4-0, (*b*) 3-1, (*c*) 2-2 among his opponents?

2.8. A poker hand contains 5 cards. A flush is a hand all of the same suit, a straight is a hand in rank order (Aces counting high or low), and a straight flush is a hand all of the same suit in rank order; these categories are exclusive so that a straight flush does not count as either a flush or a straight. What are the chances of dealing (*a*) a flush, (*b*) a straight, (*c*) a straight flush?

Chapter 3

RANDOM VARIABLES AND PROBABILITY DISTRIBUTIONS

So far we have been considering the probabilities of quite arbitrary events. Very often, however, the events in which we are interested are numerical. Such a numerical variable which takes different values with different probabilities is called a *random variable.* There are two types of random variable: *discrete* variables, such as the number of petals on a flower, which arise from counting and which can in consequence only take the integral values 0, 1, 2, ...; and *continuous* variables, such as the length of a petal or the weight of a man, which result from measuring something and can therefore take any value within a certain range. We will now discuss them in turn.

Discrete Random Variables

As an example of a discrete random variable, consider the data in Table 8 on the sizes of 815 consecutive litters of rats. We can imagine that if more and more litters of the same species were counted under the same conditions the relative frequencies in the last row would tend to stable limiting values or probabilities. Litter size may thus be thought of as taking different values with different probabilities, that is to say as a random variable.

Table 8 is said to represent the *frequency distribution* of litter size since it shows with what frequencies litters of different sizes are distributed over the possible values of litter size.

Table 8

Frequency distribution of litter size in rats (King, 1924)

Litter size	1	2	3	4	5	6	7	8	9	10	11	12	Total
No. of litters	7	33	58	116	125	126	121	107	56	37	25	4	815
Relative frequency	·01	·04	·07	·14	·15	·15	·15	·13	·07	·05	·03	·01	1·00

The data can be represented graphically in a diagram like Fig. 3. As the number of observations increases the frequency distribution will tend to a limiting *probability distribution* which will show the probabilities with which litters of different sizes are distributed over the possible values of litter size; we are in

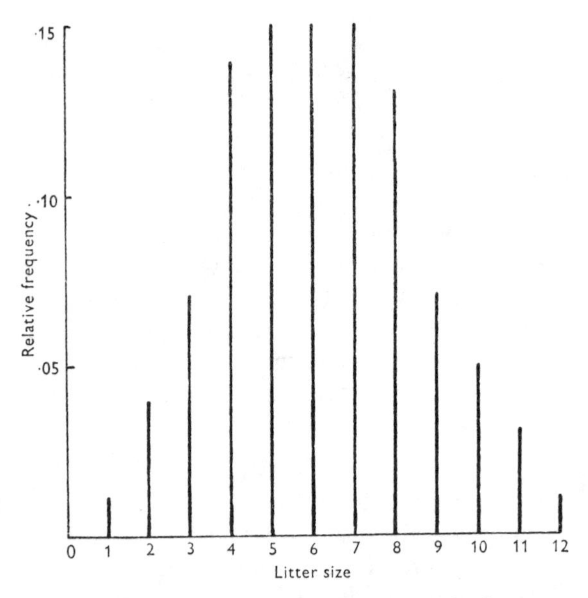

FIG. 3. Frequency distribution of litter size in rats

fact only interested in the actual frequency distribution in Table 8 because of the information which can be inferred from it about the underlying probability distribution. An example of a theoretical probability distribution is provided in Table 4 on p. 20, in which are shown the probabilities of throwing different numbers between 2 and 12 with two unbiased dice.

We must now introduce an important principle of notation. A random variable will be denoted by a capital letter, such as X, usually in the later part of the alphabet. Associated with any discrete random variable there will be a corresponding *probability function* which tells us the probability with which X takes any particular value. For example, if X is the sum of the numbers on two unbiased dice, then

$$Prob\ [X = 2] = \frac{1}{36}$$

$$Prob\ [X = 3] = \frac{2}{36}$$

and so on. This particular probability function is most easily displayed in a table, as in Table 4 (p. 20); it can however also be given in a formula

$$Prob\ [X = x] = \frac{6-|x-7|}{36}$$

$$x = 2, 3, \ldots, 12.$$

($|x-7|$ means the modulus or absolute value of $x-7$, the sign being disregarded.) We have been forced to introduce another symbol, x, to stand for a particular, but arbitrary, value that the random variable can take; once x has been

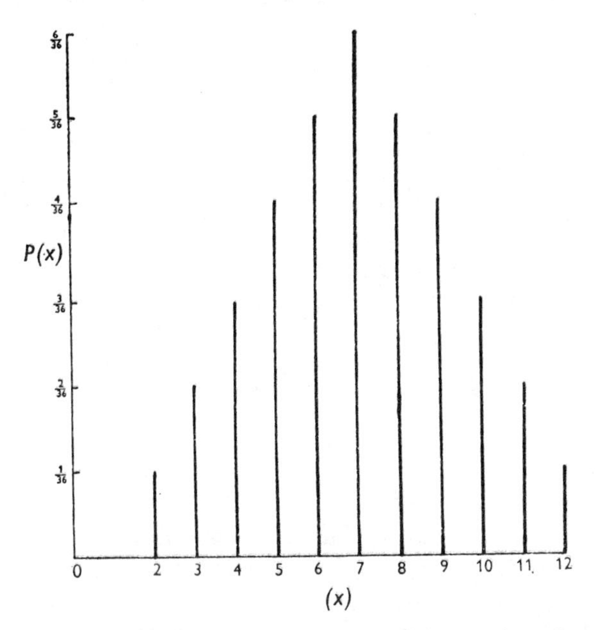

FIG. 4a. The probability function of the number of points on two dice

specified the probability that X will take that value can be calculated (provided of course that the probability function is known). The probability function is thus a function of this associated mathematical variable x: a function of the random variable itself, such as X^2 or $3X$, would on the other hand be a new random variable and not a probability function; X^2 for example would be a random variable taking the values 4, 9, 16, 25, ..., 144 with probabilities 1/36, 2/36, 3/36, 4/36, ..., 1/36. A probability function is usually denoted by $P(x)$.

Instead of the probability function, $P(x)$, it is sometimes convenient to consider the *cumulative probability function*, which specifies the probability that X is less than or equal to some particular value x and which may be denoted by $F(x)$:

$$F(x) = Prob\ [X \leq x].$$

The cumulative probability function can clearly be calculated by summing the probabilities of all values less than or equal to x:

$$F(x) = P(0)+P(1)+\ldots+P(x) = \sum_{u \leq x} P(u).$$

(The symbol Σ stands for summation and the expression on the right hand side of the above identity means the sum of the terms $P(u)$ over values of u which are less than or equal to x; it has been necessary to introduce the new symbol u because we are now thinking of x as an arbitrary constant.) If X is the sum of the numbers on two unbiased dice, for example, then

$$F(2) = \frac{1}{36}$$

$$F(3) = \frac{3}{36}$$

and so on. $F(x)$ is also defined for non-integral values of x; for example, $F(3\frac{1}{2}) = 3/36$, since the probability that X will be less than or equal to $3\frac{1}{2}$ is the same as the probability that it will be less than or equal to 3. Figs. 4a and 4b are graphical representations of $P(x)$ and $F(x)$ for this random variable. It is clear

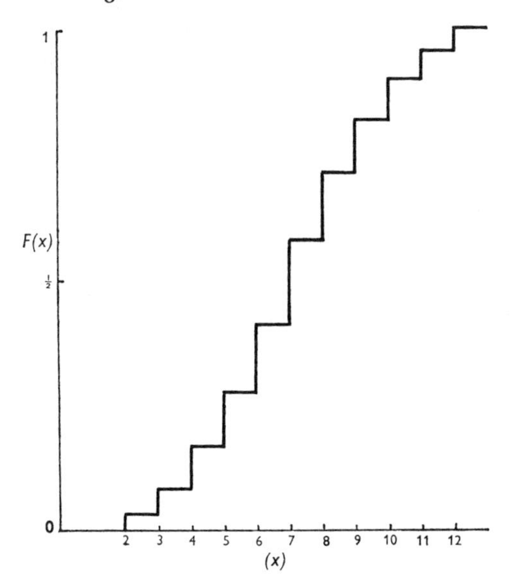

FIG. 4b. The cumulative probability function of the number of points on two dice

that for any discrete random variable $F(x)$ will be a step function increasing from zero to one and taking a jump of size $P(x)$ at each integral value of x.

CONTINUOUS RANDOM VARIABLES

So far we have been considering random variables which resulted from counting something and could therefore only take the discrete values 0, 1, 2, ... We turn now to continuous variables, which result from measurement and can therefore take any value within a certain range. Consider as an example the frequency distribution of the weight of 338 Anglo-Saxon silver pennies shown in Table 9. Each coin was weighed to the nearest tenth of a grain; the entry ' $14\frac{1}{2}$ — 18 ' therefore means that 18 coins weighed between 14·45 and 14·95 grains. It should be noted that the width of the class intervals is 1 grain in the tails of the distribution but has been reduced to $\frac{1}{2}$ grain in the centre where more observations are available.

The frequency distribution of a continuous variable like this is best represented graphically in a *histogram* as in Fig. 5. The principle of the histogram is that the *area* of each rectangle represents the proportion of observations falling in that

TABLE 9

Frequency distribution of the weight in grains of 338 silver pennies of the 'pointed helmet' type of Cnut. (Source: Butler, 1961)

Weight (grains)	Number of pennies
10 –	1
11 –	5
12 –	8
13 –	22
14 –	15
$14\frac{1}{2}$ –	18
15 –	60
$15\frac{1}{2}$ –	65
16 –	44
$16\frac{1}{2}$ –	40
17 –	22
$17\frac{1}{2}$ –	26
18 –	12
19 and over	0

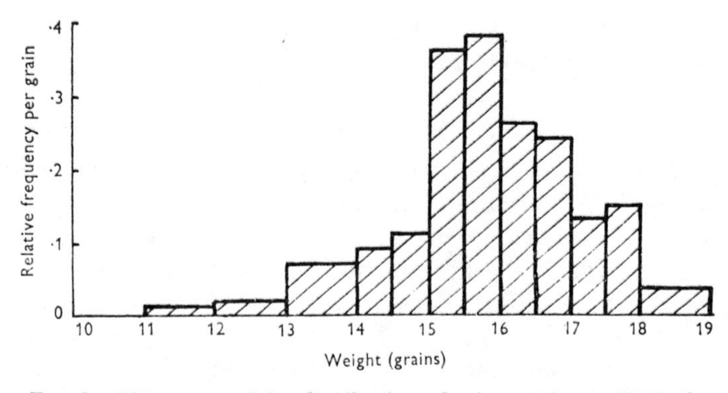

FIG. 5. Histogram of the distribution of coin weights in Table 9

interval. For example, the proportion of coins weighing between 14·45 and 14·95 grains is 18/338 = ·053; this is represented by a rectangle of the same area, that is to say with a width of $\frac{1}{2}$ and a height of $2 \times \cdot 053 = \cdot 106$, erected on the base 14·45-14·95. The ordinate in a histogram is thus not relative frequency but relative frequency per unit, in this instance per grain. The reason for representing the data in this way is that the width of the class interval is quite arbitrary and one wants to obtain roughly the same picture whatever it is and whether or not it is the same throughout the distribution. In Table 9, for example, it is obvious that the drop from 22 coins in the class 13— to 15 coins in the class 14— is due to the fact that the width of the class interval has been reduced from 1 grain to $\frac{1}{2}$ grain at this point. It would clearly be misleading to use the frequencies as they stand without making an adjustment for the difference in the class width. The histogram is therefore constructed so that the area rather than the height of each rectangle represents the relative frequency of observations in that interval. It follows that the area of the histogram between any two points represents the relative frequency of observations between those two values. In particular the area of the whole histogram is unity.

We can imagine that, if the number of observations is increased indefinitely and at the same time the class interval is made smaller and smaller, the histogram will tend to a smooth continuous curve; this curve is called the *probability density function* which we may denote by $f(x)$. The area under the density function between any two points, x_1 and x_2, that is to say the integral of the function between them, represents the probability that the random variable will lie between these two values:

$$Prob\ [x_1 < X \leq x_2] = \int_{x_1}^{x_2} f(x)dx.$$

This is illustrated in Fig. 6. If dx is a very small increment in x, so small that the density function is practically constant between x and $x+dx$, then the probability that X will lie in this small interval is very nearly $f(x)dx$, which is the area of a rectangle with height $f(x)$ and width dx. $f(x)$ may therefore be thought of as representing the probability *density* at x.

A continuous probability distribution can also be represented by its cumulative probability function, $F(x)$, which, as in the discrete case, specifies the probability that X is less than or equal to x and which is the limiting form of the cumulative frequency diagram showing the proportion of observations up to a given value. For example, in Table 9 there is 1 observation less than 10·95, 6 less than 11·95 and so on; the corresponding proportions are ·003, ·018 and so on, which are plotted in Fig. 7. As the number of observations increases,

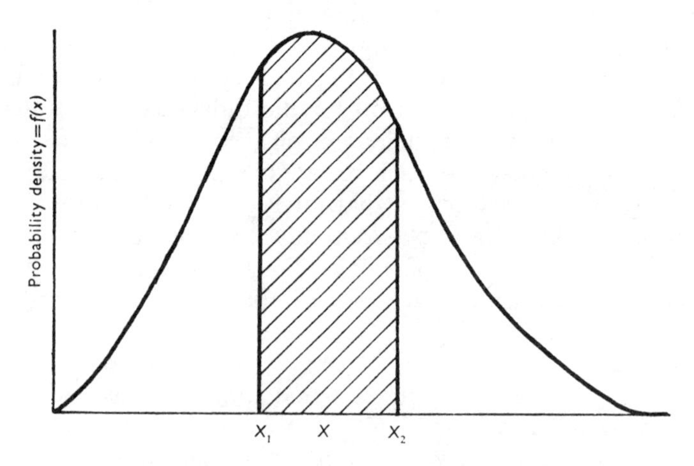

FIG. 6. The shaded area represents the probability that the random variable lies between x_1 and x_2

and the class interval is reduced so that more and more points can be plotted, the graph will tend to a limiting curve which is the cumulative probability function, $F(x)$, any point on which shows the probability that X will be less than or equal to x. As in the discrete case $F(x)$ must increase from 0 to 1 as x increases from its smallest to its largest value but it will be a smooth, continuous curve and not a step function.

It follows from what has been said in the paragraph before last that $F(x)$ is the area under the density function, that is to say the integral, to the left of and up to x:

$$F(x) = Prob\ [X \leq x] = \int_{-\infty}^{x} f(u)du.$$

Comparison with the corresponding formula for a discrete

variable on p. 32 shows that it is the same except that an integral has replaced a sum; the reason for this is that the discrete probability mass, $P(x)$, has been replaced by the infinitesimal element of probability $f(x)dx$. Conversely, the density function, $f(x)$, is the rate at which $F(x)$ is increasing, i.e. its derivative.

As a simple though rather trivial example of a continuous probability distribution we shall consider the uniform distri-

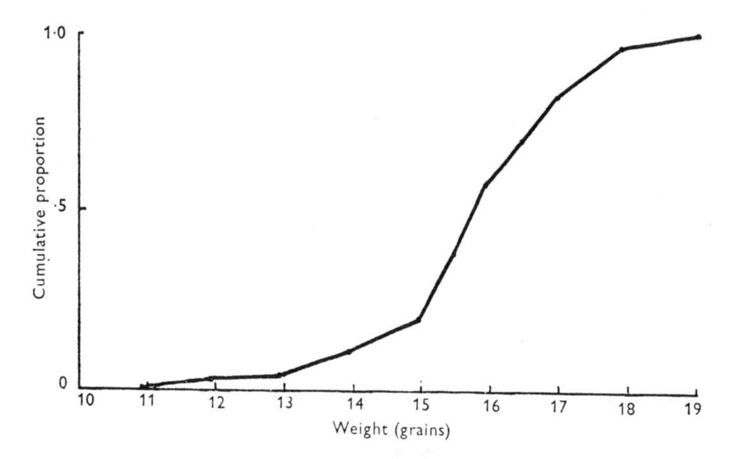

FIG. 7. Cumulative frequency diagram of the distribution of coin weights in Table 9

bution. A continuous random variable is said to be uniformly distributed between 0 and 1 if it is equally likely to lie anywhere in this interval but cannot lie outside it. For example, if we measure the height in inches of a group of men and consider the fractional part of each height ignoring the integral part, the result will be a random variable which must lie between 0 and 1 and which will probably be nearly evenly distributed over this interval. The density function of such a random variable must be a constant; and this constant must be 1 since the area under the density function must be 1. Hence

$$f(x) = 1, \; 0 \leqq x \leqq 1$$

$$f(x) = 0, \text{ otherwise.}$$

By integrating this function we find that

$$
\begin{aligned}
F(x) &= x, \ 0 \leqq x \leqq 1 \\
&= 0, \ x < 0 \\
&= 1, \ x > 1.
\end{aligned}
$$

This result is in accordance with common sense. The distribution is illustrated graphically in Fig. 8.

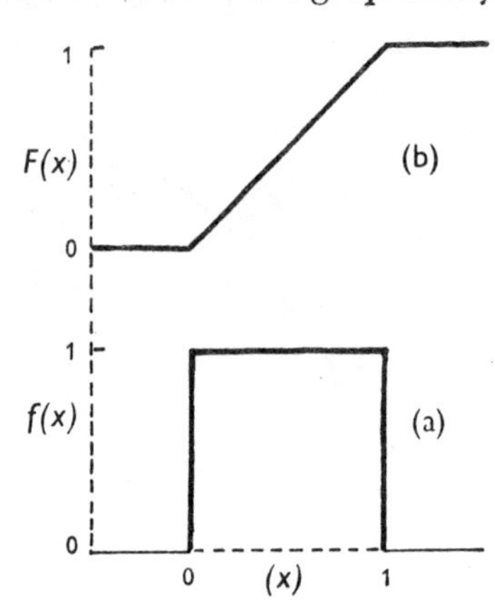

FIG. 8. (a) Probability density function and (b) cumulative probability function of the uniform distribution

MULTIVARIATE DISTRIBUTIONS

These ideas can be extended to describe the joint distribution of two or more random variables. Thus Table 3 on p. 18 shows the joint frequency distribution of the number of pips on two dice; this is called a bivariate distribution since two random variables are involved. Let us denote the numbers of pips on the red die and on the white die by X and Y respectively and let us suppose that the table has been recalculated to show relative frequencies rather than absolute frequencies and that the experiment is then continued indefinitely so that these relative frequencies tend to the corresponding probabilities. The probabilities in the main body of the table will constitute the joint probability distribution of X and Y; for example, the entry in the third row and the fourth column will be the probability that simultaneously the red die shows 3 pips and the white die 4, which we may denote by $P(3, 4)$. In general, $P(x, y)$ denotes the probability that simultaneously the red die shows x pips and the white die y pips:

$$P(x, y) = Prob\ [X = x \text{ and } Y = y].$$

This probability will be found in the xth row and the yth column of the table.

The main body of the table will thus represent the joint probability distribution of X and Y. The entries in the right hand margin, obtained by summing the probabilities in different rows, will constitute the overall probability distribution of X, regardless of the value of Y, which we may denote by $P_1(x)$:

$$P_1(x) = Prob\ [X = x] = \sum_y P(x, y).$$

Likewise the entries in the bottom margin of the table, obtained by summing the probabilities in the columns, will constitute the overall probability distribution of Y, regardless of the value of X, which we may denote by $P_2(y)$:

$$P_2(y) = Prob\ [Y = y] = \sum_x P(x, y).$$

These distributions are sometimes called the marginal distributions of X and Y respectively.

The conditional probability that X will take the value x given that Y is known to have the value y is

$$Prob\ [X = x \mid Y = y] = \frac{P(x, y)}{P_2(y)}.$$

If X and Y are statistically independent, as they are likely to be in this example, then this conditional probability distribution will not depend on y and will be equal to the overall probability distribution, $P_1(x)$, for all values of y; it follows that

$$P(x, y) = P_1(x) \times P_2(y).$$

This relationship is, of course, an expression of the law of multiplication for independent events. If, however, X and Y are not statistically independent the conditional probability distribution of X will vary with y and the above relationship will not hold.

The analysis in the case of continuous random variables is similar. Table 10, for example, shows the joint frequency distribution of head length (X) and head breadth (Y) among 3000 criminals (Macdonell, 1901). This distribution can be

TABLE 10

The joint frequency distribution of head length and head breadth in 3000 criminals (Macdonell, 1901)

Head length (centimetres)	Head breadth (centimetres)								
	13–	13½–	14–	14½–	15–	15½–	16–	16½–	Total
16–	0	0	0	0	1	0	0	0	1
16½–	0	0	1	0	1	0	0	0	2
17–	0	5	4	4	1	0	0	0	14
17½–	1	8	17	15	11	2	0	0	54
18–	0	6	55	119	74	14	1	0	269
18½–	0	5	108	264	234	75	6	1	693
19–	0	10	72	360	400	156	26	2	1026
19½–	0	1	28	174	239	160	36	7	645
20–	0	2	4	31	86	100	24	2	249
20½–	0	0	1	4	17	17	5	0	44
21–	0	0	1	0	0	1	0	1	3
Total	1	37	291	971	1064	525	98	13	3000

represented in three dimensions in a solid histogram in which each cell of the table is the base of a solid rectangular column whose volume is equal to the relative frequency of observations in the cell. We can imagine that if the number of observations were increased indefinitely and the dimensions of the cells reduced accordingly, the solid histogram would tend to a smooth surface whose height at any point (x, y) could be represented by a continuous function, $f(x, y)$; this function is the bivariate probability density function. If dx and dy are small increments in x and y respectively then $f(x, y)dxdy$ is the probability that the head length lies between x and $x+dx$ and that simultaneously the breadth of the head lies between y and $y+dy$:

$$f(x, y)dxdy = Prob\ [x<X\leqq x+dx, y<Y\leqq y+dy].$$

The marginal frequency distribution of head length, regardless of the breadth of the head, is found from the right hand margin of the table; the marginal density function, $f_1(x)$, which is the analogue of $P_1(x)$, is clearly found by integrating $f(x, y)$ over y:

$$f_1(x) = \int f(x, y)dy.$$

Similarly the marginal frequency distribution of head breadth, regardless of the length of the head, is found from the bottom margin of the table; the density function, $f_2(y)$, can be found by integrating the joint density function over x.

The conditional frequency distribution of head length among criminals with a particular breadth of head is found from the column in the main body of the table corresponding to that head breadth. Likewise the conditional probability density function of head length for a fixed head breadth is a vertical section of the bivariate density surface and is thus proportional to the bivariate density function, $f(x, y)$, with y held constant at the required value; to make the area under the curve unity it must be divided by $f_2(y)$, which is the integral of $f(x, y)$ for fixed y. The conditional probability density function of head length for fixed head breadth is therefore:

$$\frac{f(x, y)}{f_2(y)}.$$

If head length and head breadth were statistically independent this conditional distribution would be independent of the head breadth, y, and would be equal to the marginal density function, $f_1(x)$, whatever the head breadth was; it would follow in these circumstances that

$$f(x, y) = f_1(x) \times f_2(y).$$

In fact, however, it is clear that head length and head breadth are not independent but that men with long heads have broader heads than men with short heads (see Exercise 3.3); in consequence the above relationship does not hold.

This analysis can obviously be generalised to the joint distribution of any number of random variables, although pictorial representation must be abandoned.

Exercises

3.1. Plot the probability function and the cumulative probability function of the distribution of the number of points on three dice evaluated in Exercise 2.3.

3.2. If the digits in a table of random numbers are arranged in groups of four and a decimal point is placed in front of each group, the resulting variable will be a random variable following very nearly a uniform distribution between 0 and 1. Construct a sample of 50 observations from a uniform distribution by using a table of random numbers in this way, draw the histogram and the cumulative frequency diagram and compare these diagrams with the theoretical curves. Keep the data for use in Exercises 4.1 and 4.8.

3.3. Draw the histogram of the distribution of head breadth in Table 10 on p. 40, using centimetre class widths at the tails. Also draw the histograms of the distribution of head breadth for those with head lengths less than 19 cm, between 19 and 20 cm, and greater than 20 cm.

3.4. Suppose that X is a random variable taking the values -1, 0 and 1 with equal probabilities and that $Y = X^2$. Find the joint distribution and the marginal distributions of X and Y and also the conditional distribution of X given (*a*) $Y = 0$, (*b*) $Y = 1$.

3.5. If X and Y are independently and uniformly distributed between 0 and 1, what does their bivariate density function look like? Hence find the probability that Y/X is less than $\frac{1}{2}$ by considering the area of the triangle in the unit square defined by the relationship $y \leqslant \frac{1}{2}x$.

Problems

3.1. Suppose that X is a continuous random variable and that Y is a linear function of X, $Y = a + bX$ where b is positive. Denote the probability density functions of X and Y by $f(x)$ and $g(y)$ and the corresponding cumulative probability functions by $F(x)$ and $G(y)$ respectively. Then

$$G(y) = Prob[Y \leqslant y] = Prob[a + bX \leqslant y] = Prob[X \leqslant (y-a)/b]$$

$$= F\left(\frac{y-a}{b}\right)$$

and so

$$g(y) = \frac{dG(y)}{dy} = \frac{1}{b} f\left(\frac{y-a}{b}\right).$$

This method can be generalised to find the distribution of any monotonically increasing function of X. Suppose that X is uniformly distributed between 0 and 1 and find the distribution of (*a*) X^2, (*b*) $\sqrt{X}$.

3.2. (*a*) Suppose that $Y = a + bX$ where b is negative; find the distribution of Y in a similar way. (*b*) This method can be generalised to find the distribution of any monotonically decreasing function of X. Suppose that X is uniformly distributed between 0 and 1 and find the distribution of $-\log_e X$.

3.3. This method can be extended to deal with simple non-monotonic functions. If $Y = X^2$, where X can take positive or negative values, show that $G(y) = Prob[-\sqrt{y} \leqslant X \leqslant \sqrt{y}]$ and hence find $g(y)$. Find the distribution of X^2 if X is uniformly distributed between -1 and $+2$.

3.4. Suppose that a machine gun is mounted at a distance b from an infinitely long straight wall. If the angle of fire measured from the perpendicular from the gun to the wall is equally likely to be anywhere between $-\pi/2$ and $+\pi/2$ and if X is the distance of a bullet from the point on the wall opposite the gun, show that X follows the Cauchy distribution

$$f(x) = \frac{b}{\pi(b^2 + x^2)} \qquad -\infty < x < \infty.$$

3.5. Suppose that X and Y are independent, continuous random variables and that $U = X + Y$. Denote their probability density functions by $f(x)$, $g(y)$ and $h(u)$ and the corresponding cumulative probability functions by $F(x)$, $G(y)$ and $H(u)$ respectively. Then

$$H(u) = \text{Prob}\,[U \leqslant u] = \text{Prob}\,[X + Y \leqslant u] = \text{Prob}\,[X \leqslant u - Y].$$

For a fixed value of Y, say $Y = y$, this probability is $F(u-y)$, and the probability that Y will lie in the range y to $y + dy$ is $g(y)dy$. Hence the probability that $U \leqslant u$ and that simultaneously Y lies between y and $y + dy$ is $F(u-y)g(y)dy$ and so the total probability that $U \leqslant u$ is

$$H(u) = \int F(u-y)g(y)dy,$$

whence

$$h(u) = \int f(u-y)g(y)dy,$$

provided that the range of integration is independent of u.

Suppose that X and Y are independently uniformly distributed between 0 and 1. Find the distribution of $U = X + Y$. [Hint: in finding $H(u)$ consider separately the two cases $0 \leqslant u \leqslant 1$ and $1 \leqslant u \leqslant 2$. Be careful.]

3.6. Suppose that X and Y are independent, continuous random variables, that Y is essentially positive and that $U = X/Y$. Then, in a notation analogous to that of Problem 3.5,

$$H(u) = \text{Prob}\,[U \leqslant u] = \text{Prob}\,[X/Y \leqslant u] = \text{Prob}\,[X \leqslant uY]$$

By an argument similar to that of Problem 3.5 it follows that

$$H(u) = \int F(uy)g(y)dy,$$

whence

$$h(u) = \int f(uy)y\,g(y)dy,$$

provided that the range of integration is independent of u.

Suppose that X and Y are independently uniformly distributed between 0 and 1. Find the distribution of $U = X/Y$. [Hint: consider separately the two cases $0 \leqslant u \leqslant 1$ and $1 \leqslant u < \infty$.]

3.7. The master of a ship can fix its position by taking bearings of two known points on land. However, if three bearings are taken the position lines will usually not meet in a point because of errors of measurement but will form a triangular 'cocked hat' in which it is assumed that the ship lies. If the observed bearings are symmetrically and independently distributed about the true bearings show that the probability that the ship lies in the cocked hat is $\frac{1}{4}$.

3.8. I am to set out for Omega tomorrow morning at 9.30, and find that I am indifferent between two routes, in the sense that I am equally likely to be late for an appointment at 3.15.

If I take the first route I will get to Beta between noon and 1.30. It normally takes a quarter of an hour to get through Beta, but one's speed is reduced by two thirds between 1.00 and 2.00. From the far side to Omega takes two hours.

The second route is through open country, so that one can be sure of averaging between forty and fifty miles per hour.

How many miles is the second route?

NOTE: All the probability distributions mentioned are uniform. [Certificate, 1965.]

Chapter 4

DESCRIPTIVE PROPERTIES OF DISTRIBUTIONS

It is often useful to summarise the main properties of a distribution, such as its centre, its variability and its general shape, in one or two descriptive measures; in this chapter we shall consider how this can best be done. We shall begin by discussing measures of the centre or location of a distribution.

Measures of Location

There are three commonly used measures of the centre or location of a distribution; they are, in order of importance, the mean, the median and the mode.

Mean. The mean or average of a set of observations is simply their sum divided by their number. For example, the mean of the five observations

$$11, 12, 13, 9, 13$$

is $58/5 = 11{\cdot}6$ since they add up to 58. The general formula for the mean, $\bar{x}$, of n observations, $x_1, x_2, \ldots, x_n$, is

$$\bar{x} = \frac{x_1+x_2+\ldots x_n}{n} = \sum_{i=1}^{n} x_i/n.$$

If, on the other hand, the data are given in the form of a frequency distribution in which we are told that the value x has occurred $n(x)$ times, then the sum of the observations is $\sum_x xn(x)$ so that the formula for the mean is

$$\bar{x} = \sum_x xn(x)/n.$$

For example, to find the mean litter size for the data in Table 8 on p. 29, we first calculate the total number of offspring in all the litters, which is

$$1\times 7+2\times 33+\ldots+12\times 4 = 4992$$

and then divide this by 815, the number of litters, to obtain the mean or average litter size of 4992/815 = 6·125. To perform the corresponding calculation for the grouped frequency distribution of a continuous variable it is assumed that all the observations in a class interval are concentrated at the middle of that interval. For example, the mean weight of the 338 coins whose weight distribution is shown in Table 9 on p. 34 is calculated as

$$\frac{1\times 10{\cdot}45+5\times 11{\cdot}45+\ldots+26\times 17{\cdot}70+12\times 18{\cdot}45}{338}$$
$$= 15{\cdot}722 \text{ grains.}$$

(It should be noted that 10·45 is the centre of the class interval 10— since it contains coins weighing between 9·95 and 10·95 grains; similarly 17·70 is the centre of the class interval $17\frac{1}{2}$— since it contains coins weighing between 17·45 and 17·95 grains.)

If we write $p(x) = n(x)/n$ for the proportion of times on which the value x has occurred, then the formula for the mean can be re-written as

$$\bar{x} = \sum_x xp(x).$$

That is to say, the mean is the sum of the possible values which the variable can take, each value being multiplied by its relative frequency of occurrence. If we are considering a discrete distribution then the proportions, $p(x)$, will, as the number of observations increases, tend to the corresponding probabilities, $P(x)$, and so $\bar{x}$ will tend to the limiting value

$$\sum_x xP(x)$$

which is the sum of the possible values which the variable can take, each value being multiplied by its probability of occurrence. This quantity can therefore be taken as a measure of the centre of a probability distribution. It is known as the mean of the probability distribution and is usually denoted by μ. For example, the mean of the distribution of the sum of the points on two dice (see Table 4 on p. 20) is

$$2\times\frac{1}{36}+3\times\frac{2}{36}+\ldots+12\times\frac{1}{36} = 7.$$

In the case of a continuous distribution the probability mass $P(x)$ is replaced by the infinitesimal element of probability $f(x)dx$; the mean of the distribution towards which $\bar{x}$ tends as the number of observations increases is therefore given by the integral

$$\int xf(x)dx$$

evaluated over the possible range of values of the variable. For example, if X is uniformly distributed between 0 and 1, then $f(x)$ is 1 between 0 and 1 and zero elsewhere; the mean of the distribution is therefore

$$\int_0^1 xdx = [\tfrac{1}{2}x^2]_0^1 = \tfrac{1}{2}.$$

It is important to distinguish carefully between the mean, $\bar{x}$, of a set of observations or a frequency distribution, and the mean, μ, of the corresponding probability distribution towards which $\bar{x}$ tends as the number of observations increases. A measure, such as the mean, $\bar{x}$, which can be calculated from an observed frequency distribution is called a *statistic*, while the corresponding measure of the probability distribution is called a *parameter*. In order to distinguish between them it is usual to denote statistics by Roman letters and parameters by Greek letters. When this is inconvenient a statistic is denoted by a small letter and a parameter by the corresponding capital letter. For example, π cannot be used for a parameter since by mathematical custom it is used for the ratio of the circumference to the diameter of a circle; consequently p is used to denote an observed proportion and P for the corresponding probability. In practice, of course, a statistic is only of interest because of the information which can be inferred from it about the corresponding parameter; the method of extracting this information constitutes the subject matter of statistical inference which will be considered later.

Median. We turn now to the second measure of the ‘ centre ’ of a distribution, the median. The median of a set of observations is the middle observation when they are arranged in order of magnitude. Thus the median of the five observations

11, 12, 13, 9, 13

is 12 since it is the third observation when they are rearranged in rank order (9, 11, 12, 13, 13). The median litter size of the data in Table 8 (p. 29) is the size of the 408th litter, which is 6 since there are 339 litters of size 5 or less and 465 of size 6 or less. When there is an even number of observations the median is defined as lying half-way between the two middle ones. To estimate the median from a grouped frequency distribution of a continuous variable we interpolate linearly in the class interval in which it lies. For example, the weights of 338 coins are recorded in Table 9 on p. 34, so that the median is half-way between the weights of the 169th and the 170th coins. Now there are 129 coins which weigh up to 15·45 grains, and 194 coins which weigh up to 15·95 grains. The median, therefore, lies somewhere in this interval and is calculated as

$$15{\cdot}45+\frac{40{\cdot}5}{65}\times\tfrac{1}{2} = 15{\cdot}7615 \text{ grains.}$$

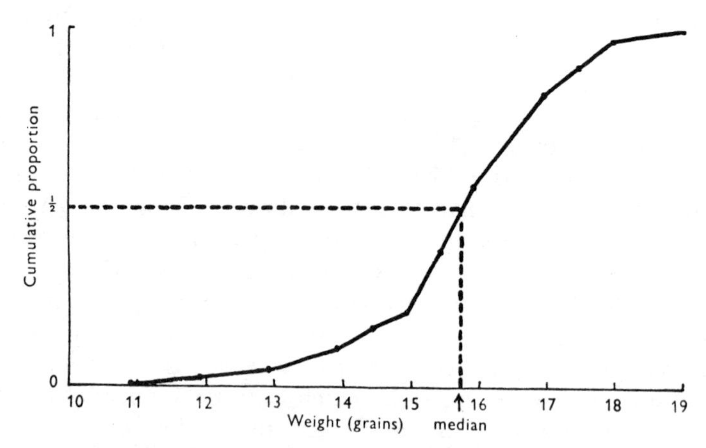

FIG. 9. Graphical representation of the median of the distribution of coin weights (compare Fig. on p. 37)

Graphically, the median is the point at which the cumulative frequency function reaches $\frac{1}{2}$; this is illustrated in Fig. 9. As the number of observations increases the cumulative frequency function tends towards the cumulative probability function, $F(x)$; the median of the probability distribution, towards which the observed median tends, is therefore the point at which $F(x)$ first reaches $\frac{1}{2}$; this is illustrated in Fig. 10.

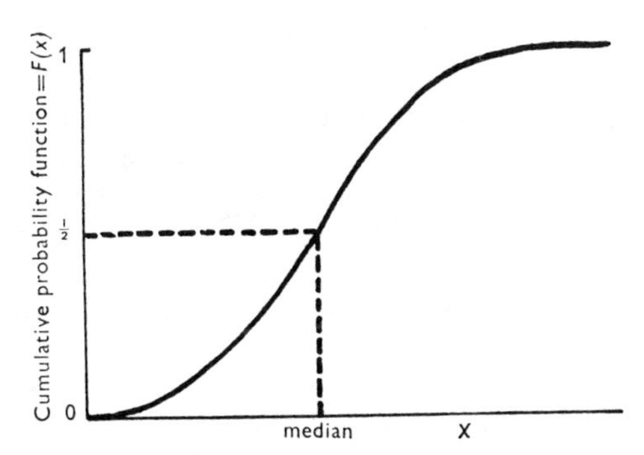

FIG. 10. Graphical representation of the median of a probability distribution

The median of a continuous probability distribution divides the distribution in two equal parts in the sense that

$$Prob\ [X<\text{median}] = Prob\ [X>\text{median}] = \tfrac{1}{2}.$$

This is not true for a discrete distribution, because a non-zero probability is concentrated at the median itself. For example, the median of the distribution of the sum of points on two dice (p. 20) is 7; in this case

$$\begin{aligned} Prob\ [X<7] &= 15/36 \\ Prob\ [X=7] &= \ \ 6/36 \\ Prob\ [X>7] &= 15/36. \end{aligned}$$

It is true that there is an equal probability that the random variable will be above or below the median, but even this will not generally be so. Thus we might equally well have a probability of 13/36 that X will be less than 7, a probability of 6/36 that it will be equal to 7 and a probability of 17/36 that it will be greater than 7. The median would still be 7, but the random variable would be more likely to lie above this value than below it. For this reason the median is more useful in describing a continuous than a discrete distribution.

Mode. The third measure of location is the mode, which is the value of the most frequent observation. The mode of a discrete probability distribution is the value which has the highest probability of occurring; thus the mode of the distribution of the sum of points on two dice shown in Table 4 on p. 20 is 7. Similarly the mode of a continuous probability distribution is the point at which the density function attains its highest value and can be found by solving the equation

$$\frac{df(x)}{dx} = 0.$$

(We are assuming for the moment that the curve has only one maximum.)

There is an interesting empirical relationship between the mean, the median and the mode of a continuous probability distribution. If the distribution is symmetrical they will, of course, all be the same. If, however, it is skew, the mean will lie nearer the long tail of the distribution than the median and the mode will lie on the other side of the median from the mean. That is to say, the mean, the median and the mode will either occur in that order (their alphabetical order) or in the reverse order, depending on whether the distribution is skew to the left or the right. Furthermore, the distance between the mode and the median is usually about twice the distance between the median and the mean. This rather remarkable relationship is illustrated in Fig. 11.

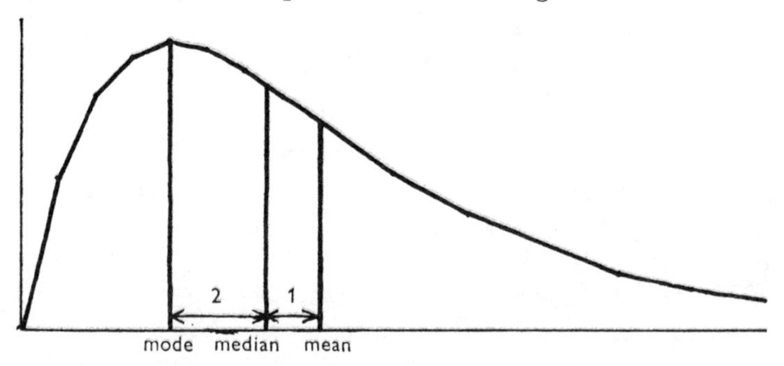

FIG. 11. Relationship between the mean, the median and the mode of a typical continuous probability distribution

The mode is a useful quantity in describing the properties of probability distribution but it is rather doubtful whether it can be regarded as a measure of the 'centre' of the distribution. Consider, for example, the exponential distribution, which is discussed in more detail in Chapter 6 and whose density function is

$$f(x) = \lambda e^{-\lambda x}, \; 0 \leqq x < \infty.$$

The mode of the distribution is 0, but this is clearly not a reasonable measure of its 'centre' since the variable must be positive.

The mode is seldom used as a statistic for describing a frequency distribution because of the difficulty of defining it. In Fig. 5 on p. 34, for example, it is clear that the modal value lies somewhere between 15 and 16 grains, but it is difficult to see how it could be located more precisely without drawing a theoretical curve through the histogram; this procedure would be to some extent arbitrary because the position of the mode would depend on the form of the curve which had been chosen to represent the data. In a less regular histogram based on fewer observations the problem becomes even more acute. It has also been suggested that the mode can be calculated from the observed mean and median by the formula

$$\text{mode} = \text{mean} + 3(\text{median} - \text{mean}).$$

This formula assumes the truth of the relationship between these three quantities described above and is thus again begging the question. In the case of a discrete frequency distribution the mode can be defined without ambiguity as the value of the most frequent observation, but this has little meaning unless there is a considerable number of observations, and even then difficulties arise if there appears to be more than one modal value. For these reasons the mode is not used as a precisely defined statistic characterising a frequency distribution, although inspection of the data will usually enable a fairly good estimate to be made by eye of the mode of the corresponding probability distribution.

The mean v. the median

We must now compare the merits of the mean and the median as measures of the 'centre' of a distribution. We shall confine our attention to continuous distributions since we have seen that the median is rather unsuitable for describing discrete distributions. Suppose first that the underlying probability distribution is symmetrical. In this case the mean and the median of the distribution will be the same and will coincide with the centre of symmetry; this must clearly be true of any sensible measure of location. If, however, we take n observations from this distribution and calculate their mean and median, it is very unlikely that they will be identical although they will both be estimates of the same parameter. The question to be considered, therefore, is whether the sample mean or median provides, on the average, the more accurate estimate of this parameter. This question will be considered in detail in Chapter 11. The answer depends on the shape of the underlying distribution, but it turns out that the sample mean is usually, though not always, more accurate than the median. If we are prepared to assume that the underlying distribution is symmetrical, therefore, we shall probably do better by employing the sample mean rather than the median to estimate its centre.

Let us now suppose that the underlying probability distribution is skew, so that one tail of the distribution extends further out than the other. In this case the mean and the median of the distribution will not be the same and the mean will lie further out on its long tail than the median (see Fig. 11 on p. 50). Suppose, for example, that Fig. 11 represents a distribution of incomes which will have a long tail extending out to the right because of the very small number of people with very large incomes. The mean will be larger than the median because it is affected by the actual values of the very large incomes, whereas the median is only affected by the very small number of people with such incomes. The sample mean and median will therefore be measures of different parameters and in deciding which of them to use we have to decide which of these parameters of the probability distribution we wish to estimate. The mean of a probability distribution is its centre of gravity, while the median divides it into two equal parts. Sometimes one and sometimes the other of these two measures will suit our purposes better. For example, the median might be a more appropriate measure than the mean of the income of the 'typical man', whereas the mean might be more appropriate if we were comparing average incomes in England and Scotland as a measure of economic prosperity.

The choice between the mean and the median as measures of the 'centre' of an asymmetric probability distribution must remain to some extent arbitrary. There are, however, two further considerations to guide us. The first is that the median is invariant under monotonic transformations of the random variable. For example, if X is an essentially positive variable with median $\tilde{\mu}$, then the median of $\sqrt{X}$ will be $\sqrt{\tilde{\mu}}$ since, if the probability that X is less than $\tilde{\mu}$ is $\frac{1}{2}$, then the probability that $\sqrt{X}$ is less than $\sqrt{\tilde{\mu}}$ must also be $\frac{1}{2}$. The same must hold for any monotonic function of X, that is to say any function of X which either always increases or always decreases as X increases. The mean of a probability distribution, however, is not invariant, except under linear transformations. Perhaps an example will make this point clearer. Consider three squares with sides of 3, 4 and 5 inches and consequently with areas of 9, 16 and 25 square inches. The median length of the sides is 4 inches and the median area of the squares is 16 square inches, which is of course the square of 4. The mean length of the sides is also 4 inches, but the mean area of the squares is $(9+16+25)/3 = 16{\cdot}67$ square inches.

The median is invariant because it is calculated simply from ranking the observations and does not involve any arithmetical operations on them. This property provides a strong argument for the use of the median when the scale of measurement is not fully quantitative, as may happen in the social sciences. For example, a psychologist might construct a scale of introversion/extroversion between 0 and 10. If Brown, Jones and Smith have scores on this scale of 3, 4 and 6, the psychologist would conclude that Smith was more extroverted than Jones and Jones than Brown, but he might well be unwilling to conclude that Smith was twice as extroverted as Brown, or even that the difference in the degree of extroversion between Smith and Jones was twice as large as the difference between Jones and Brown. Under these circumstances it is to some extent arbitrary whether the original scores, or their squares, or square roots, or logarithms, are used to 'measure' extroversion and the median would be a more appropriate measure of the 'centre' of the distribution (see Siegel, 1956).

The second consideration to guide us in choosing between the mean and the median as measures of location is one of mathematical convenience. In statistical theory we often wish to study the properties of sums of random variables; thus if X and Y are random variables, their sum, $X+Y$, will be a new random variable whose probability distribution can be determined if we know the distributions of X and Y. Now it happens that the mean of the sum of two or more random variables is equal to the sum of their means (a full discussion will be given in the next chapter). No such simple relation holds between the medians of the corresponding distributions. It follows that the mean of a distribution can often be found quite easily, whereas its median is much less tractable. This is undoubtedly the main reason why statisticians usually choose the mean rather than the median to describe the 'centre' of a probability distribution.

Measures of Dispersion

The characteristic feature of a random variable is its variability, and it is clearly important to be able to measure the amount of this variability. Perhaps the most obvious measure of this quantity is the *mean deviation* which is the average of the absolute deviations from the mean. For example, we have seen that the mean of the five observations

$$11, 12, 13, 9, 13$$

is 11·6; the deviations from the mean, regardless of sign, are

$$0{\cdot}6,\ 0{\cdot}4,\ 1{\cdot}4,\ 2{\cdot}6,\ 1{\cdot}4$$

whose average is $6{\cdot}4/5 = 1{\cdot}28$. As the number of observations increases, the mean deviation tends towards a limiting value, which is the mean deviation of the corresponding probability distribution, given by the formula

$$\text{mean deviation} \begin{cases} = \sum_x |x-\mu| P(x) & \text{in the discrete case} \\ = \int |x-\mu| f(x)dx & \text{in the continuous case.} \end{cases}$$

For example, the mean deviation of the uniform distribution is

$$\int_0^1 \mid x-\tfrac{1}{2} \mid dx = 2\int_{\frac{1}{2}}^1 (x-\tfrac{1}{2})dx = 2[\tfrac{1}{2}x^2-\tfrac{1}{2}x]_{\frac{1}{2}}^1 = \tfrac{1}{4}.$$

Another measure of variability is the *interquartile range.* Just as the median is the middle observation when they are arranged in rank order, so the upper quartile is the observation lying three-quarters of the way, and the lower quartile is the observation one-quarter of the way, from the smallest observation. Thus the upper and lower quartiles of the five observations in the preceding paragraph are 13 and 11 respectively. (These are the fourth and second observations in rank order; 4 is three-quarters of the way and 2 one-quarter of the way between 1 and 5.) The interquartile range is $13-11 = 2$.

Graphically the lower and upper quartiles are the points at which the cumulative frequency function first reaches the values of $\frac{1}{4}$ and $\frac{3}{4}$ respectively. As the number of observations increases the cumulative frequency function tends to the corresponding cumulative probability function, $F(x)$, and so the lower and upper quartiles of the probability distribution, towards which the observed quartiles tend, are the points at which $F(x)$ first reaches the values of $\frac{1}{4}$ and $\frac{3}{4}$. If the distribution is continuous they are the solutions of the equations

$$F(x) = \tfrac{1}{4}$$
$$\text{and } F(x) = \tfrac{3}{4}.$$

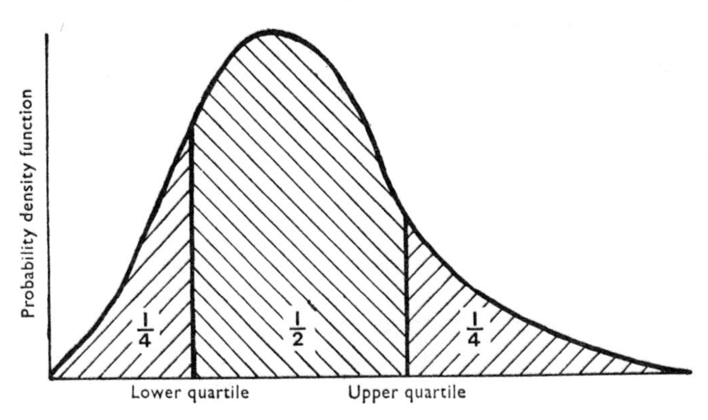

FIG. 12. The quartiles of a continuous probability distribution

Thus for the uniform distribution $F(x) = x$, provided that x lies between 0 and 1; the quartiles of the distribution are therefore $\frac{1}{4}$ and $\frac{3}{4}$, and the interquartile range is $\frac{1}{2}$. In the continuous case there is a probability of $\frac{1}{4}$ that the random variable will lie below the lower quartile, a probability of $\frac{1}{4}$ that it will lie above the upper quartile, and a probability of $\frac{1}{2}$ that it will lie between the two quartiles. This is illustrated in Fig. 12. These statements will not be exactly true for a discrete distribution. The interquartile range is thus a more useful measure of dispersion in the continuous than in the discrete case.

In practice, however, statisticians almost invariably use a third measure of dispersion, the *standard deviation*. The standard deviation of a set of observations, or a frequency distribution, is the square root of the average of the squared deviations from the mean. For example, the squared deviations from the mean of the five observations given in the first paragraph of this section are

$$0{\cdot}36,\ 0{\cdot}16,\ 1{\cdot}96,\ 6{\cdot}76,\ 1{\cdot}96$$

whose average is $11{\cdot}20/5 = 2{\cdot}24$. This quantity is called the *variance* of the observations, denoted by m_2; its square root, the standard deviation, is 1·50. It will be noticed that the variance must be positive since all the squared deviations are positive. Furthermore, the larger these deviations, the greater will the variance be; it can therefore be used as the basis for a measure of variability. Its square root is used in order to have a quantity of the same dimensions as the observations. For example, if the measurements are in feet, then the variance is in square feet but the standard deviation is in feet.

The variance of a set of observations, or a frequency distribution, is the sum of the squared deviations from the mean, S^2, divided by the number of observations:

$$m_2 = \frac{S^2}{n}.$$

If the data consist of the n observations $x_1, x_2, \ldots, x_n$, then the sum of the squared deviations from the mean is

$$S^2 = \sum_{i=1}^{n} (x_i - \bar{x})^2.$$

If the data are given in the form of a frequency distribution in which we are told that the value x has occurred $n(x)$ times, then the formula for the sum of the squared deviations is

$$S^2 = \sum_x (x-\bar{x})^2 n(x).$$

In practice it is rather tedious to compute S^2 directly in this manner and it is usually calculated by first finding the sum of the squares of the observations themselves, that is to say $\sum_{i=1}^{n} x_i^2$ if the original observations are available or $\sum_x x^2 n(x)$ if the data are given as a frequency distribution, and then subtracting a *correction factor*, $n\bar{x}^2$. The two methods of calculating S^2 give the same answer because

$$\sum_{i=1}^{n} (x_i-\bar{x})^2 = \sum_{i=1}^{n} (x_i^2 - 2\bar{x}x_i + \bar{x}^2)$$
$$= \sum_{i=1}^{n} x_i^2 - 2\bar{x} \sum_{i=1}^{n} x_i + n\bar{x}^2 = \sum_{i=1}^{n} x_i^2 - n\bar{x}^2.$$

The variance of the frequency distribution of a discrete variable can be written in the form

$$m_2 = \sum_x (x-\bar{x})^2 p(x).$$

As the number of observation, increases the sample mean, $\bar{x}$, will tend to the population mean, μ, and the proportions, $p(x)$, will tend to the corresponding probabilities, $P(x)$; hence m_2 will tend to the limiting value

$$\mu_2 = \sum_x (x-\mu)^2 P(x)$$

which is consequently the variance of the probability distribution and which is often denoted by σ^2. For example, the variance of the distribution of the sum of the numbers on two dice given in Table 4 on p. 20 is

$$5^2 \times \tfrac{1}{36} + 4^2 \times \tfrac{2}{36} + \ldots + 5^2 \times \tfrac{1}{36} = \tfrac{210}{36} = 5\tfrac{5}{6}.$$

The standard deviation is $\sqrt{(210/36)} = 2{\cdot}415$. The variance of a continuous probability distribution is

$$\mu_2 = \int (x-\mu)^2 f(x)\,dx.$$

Thus the variance of the uniform distribution is

$$\int_0^1 (x-\tfrac{1}{2})^2 dx = [\tfrac{1}{3}(x-\tfrac{1}{2})^3]_0^1 = \tfrac{1}{12}.$$

We must now consider why the standard deviation rather than the mean deviation or the interquartile range is almost always used as a measure of variability. The reasons are very much the same as those for using the mean rather than the median as a measure of location. First, let us assume that the *shape* of the underlying probability distribution, but not its mean or variance, is known. In this case, once one of the three possible parameters measuring its variability is known, then the other two can be calculated. For example, if the random variable is known to follow a normal distribution with standard deviation σ, then its mean deviation is $\cdot798\sigma$ and its interquartile range is $1\cdot349\sigma$ (see Chapter 7). If, therefore, we wish to estimate σ from a sample of n observations from this distribution, we can either use the observed standard deviation, or the observed mean deviation divided by $\cdot798$, or the observed interquartile range divided by $1\cdot349$. These three statistics will differ slightly from each other, and from σ, because of sampling variations, but they are all estimates of the same parameter, σ. The question to be considered, therefore, is which of them provides, on the average, the most accurate estimate of this parameter; as we shall see in Chapter 11 the answer is in favour of the standard deviation. It has been assumed in this argument that the underlying probability distribution is known to be normal. In practice this will never be exactly true, but since many naturally occurring distributions are approximately normal it provides good reason for using the observed standard deviation as a measure of the variability of a frequency distribution.

There is also a sound reason of mathematical convenience for using the standard deviation as a parameter to measure the variability of a probability distribution. In discussing measures of location we stated that the mean of the sum of two random variables was equal to the sum of their means. It will be shown in the next chapter that if the two random variables are independently distributed the variance of their sum is equal to the sum of their variances. No such simple relationship holds between the mean deviations or the inter-quartile ranges of the corresponding probability distributions. It follows that the variance, and hence the standard deviation, of a random variable can often be found quite easily when the other two parameters cannot.

These two reasons together account for the almost universal use of the standard deviation as a measure of dispersion.

The Shape of Distributions

We have so far discussed how to find measures of the ' centre ' of a distribution and of the degree of dispersion about this central value. The third important respect in which distributions may differ is in their shape and we shall now consider how this rather elusive concept can be described.

Multimodal distributions

We may first distinguish between unimodal distributions, which have only one peak, and multimodal distributions, which have several peaks. If an observed frequency distribution has several peaks which are real, in the sense that they are not due to sampling fluctuations but reflect peaks in the corresponding probability distribution, it is usually a sign that the distribution is a composite one made up of several unimodal distributions. We saw in Fig. 5 (p. 34) that the distribution of the weights of the ' pointed helmet ' pennies of Cnut has a single peak at a little less than 16 grains. Fig. 13 (p. 60) shows the comparable distribution for the ' quatrefoil ' pennies of Cnut; this distribution clearly has two peaks, one in the same place as the ' pointed helmet ' pennies and a subsidiary peak at about 21 grains. Now the ' pointed helmet ' pennies were being minted during the middle of Cnut's reign (*c.* 1023-1029), whereas the ' quatrefoil ' penny was his first coin type and was minted from about 1017 to 1023. The subsidiary peak at 21 grains corresponds to the weight of coins produced at the end of the reign of his predecessor Aethelred the Unready. It thus seems likely that Cnut reduced the weight standard of the penny from 21 to 16 grains, possibly to accord with Scandinavian standards, but that he did not introduce this change until two or three years after his accession, with the result that two weight standards are represented in his first coin type.

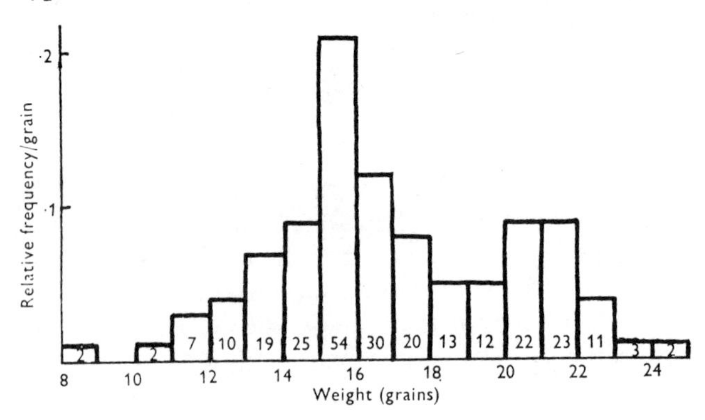

FIG. 13. Weight distribution of the 'quatrefoil' penny of Cnut (Butler, 1961)

Many biological examples of bimodal distributions are provided by quantitative characters controlled by a single pair of genes. Thus to many people the chemical substance phenylthiourea tastes bitter even in very dilute solutions, while others can only taste it in relatively concentrated solutions. This difference in the ability to taste phenylthiourea is believed to be genetically determined and to depend on a single dominant gene. The distinction between 'tasters' and 'non-tasters' is not an absolute one, but determination of the threshold of taste in large groups of people by Hartmann (1939) and Harris and Kalmus (1949) has shown a clear bimodal distribution of thresholds with a small overlap between the two classes.

As an example of a distribution with more than two modes d'Arcy Thompson in his book *On Growth and Form* has quoted the size of fish-eggs in a sample netted at the surface of the sea. The distribution shows four distinct peaks which can be attributed to the eggs of different species of fish.

A multimodal distribution thus usually represents the mixture of several different distributions. The converse of this statement is not, however, necessarily true. If two populations with unimodal distributions are mixed together, the resulting distribution will have two peaks if the two distributions are well separated; but if there is a high degree of overlap the resulting distribution may have only one rather broad peak. It is often a difficult matter to decide on such evidence whether a population is homogeneous or whether it is composed of two or more distinct sub-populations.

Skewness

Most distributions are unimodal. The main difference in shape among such distributions is in their degree of symmetry. Distributions whose right-hand tail is longer than the left-hand one are called skew to the right and *vice versa* (see Fig. 14). A numerical measure of skewness can be constructed from the average of the cubed deviations from the mean, which

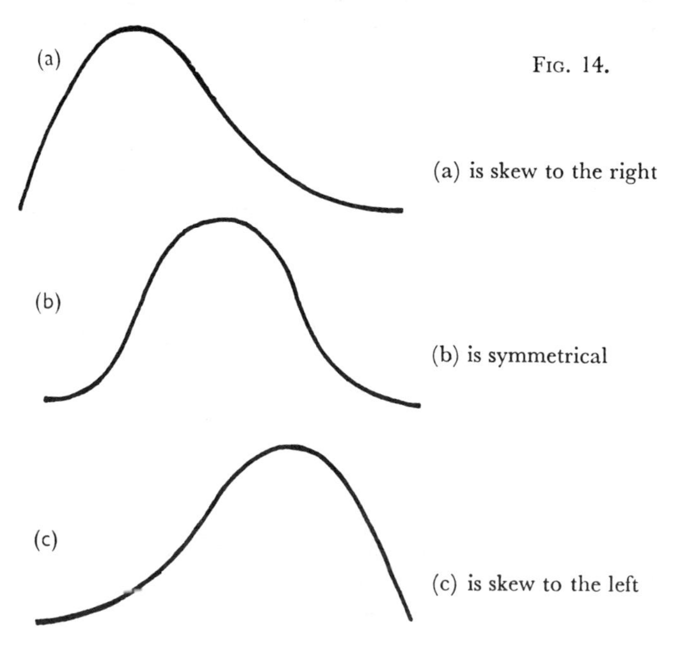

FIG. 14.

(a) is skew to the right

(b) is symmetrical

(c) is skew to the left

is called the third moment about the mean and is denoted by m_3:

$$m_3 = \sum_{i=1}^{n} (x_i - \bar{x})^3/n.$$

If the distribution is symmetrical then m_3 will be nearly zero because positive and negative deviations, when cubed, will cancel each other out. If, however, the distribution is skew to the right then m_3 will be positive because the large positive deviations, when cubed, will outweigh the negative deviations; similarly, if the distribution is skew to the left m_3 will be negative. For example, the deviations from the mean of the numbers 2, 3, 4 are −1, 0, 1, whose cubes add up to zero; on the other hand, the deviations of the numbers 2, 3, 7 from their mean are −2, −1, 3 whose cubes add up to 18.

It is clear, however, that m_3 cannot be used as it stands as a measure of skewness since it is dimensional; if the observations are expressed in inches then m_3 will be in cubic inches and so on. A measure of the shape of a distribution, on the other hand, should be dimensionless. The most convenient way to obtain such a measure is to divide m_3 by the cube of the standard deviation:

$$\text{skewness} = m_3/m_2^{3/2}.$$

This measure will not depend on the units in which the observations are expressed. A symmetrical distribution will have zero skewness; an asymmetrical distribution will have a positive or a negative measure of skewness according as it is skew to the right or to the left.

In numerical work the third moment is most easily calculated from the formula

$$m_3 = m'_3 - 3m'_2\bar{x} + 2\bar{x}^3.$$

In this formula m'_2 and m'_3 are respectively the averages of the squares and cubes of the observations themselves:

$$m'_2 = \sum_{i=1}^{n} x_i^2/n, \quad m'_3 = \sum_{i=1}^{n} x_i^3/n.$$

These quantities are called the second and third moments

about the origin. The formula follows from dividing by n both sides of the identity

$$\sum_{i=1}^{n}(x_i-\bar{x})^3 = \sum_{i=1}^{n}(x_i^3-3x_i^2\bar{x}+3x_i\bar{x}^2-\bar{x}^3)$$

$$= \sum_{i=1}^{n}x_i^3-3\bar{x}\sum_{i=1}^{n}x_i^2+2n\bar{x}^3.$$

The third moment, m_3, of a set of observations or a frequency distribution will tend, as the number of observations increases, to the third moment of the corresponding probability distribution which is denoted by μ_3 and which is given by the formula

$$\mu_3 = \sum_{x}(x-\mu)^3P(x) \text{ for a discrete distribution.}$$

$$\mu_3 = \int(x-\mu)^3 f(x)dx \text{ for a continuous distribution.}$$

The corresponding measure of skewness is μ_3/σ^3. As a rough guide we may consider a distribution with a skewness greater than 1 in absolute value as highly skew, a distribution with a skewness between $\frac{1}{2}$ and 1 as moderately skew, and a distribution with a skewness between 0 and $\frac{1}{2}$ as fairly symmetrical. The distribution in Fig. 11 (p. 50) has a skewness of 1·4. The extremely skew exponential distribution considered in Chapter 6 has a skewness of +2.

Kurtosis

The average of the fourth powers of the deviations from the mean is known as the fourth moment about the mean and denoted by m_4:

$$m_4 = \sum_{i=1}^{n}(x_i-\bar{x})^4/n.$$

This quantity is most easily calculated from the formula

$$m_4 = m'_4-4m'_3\bar{x}+6m'_2\bar{x}^2-3\bar{x}^4$$

where m'_4 denotes the average of the fourth powers of the observations themselves. This formula is derived in the same way as the corresponding formula for the third moment. As the number of observations increases, m_4 tends to a limiting value denoted by μ_4 which is the fourth moment about the mean of the corresponding probability distribution and is given by the formula

$$\mu_4 = \sum_x (x-\mu)^4 P(x) \text{ for a discrete distribution.}$$

$$\mu_4 = \int (x-\mu)^4 f(x)dx \text{ for a continuous distribution.}$$

The dimensionless quantity m_4/m_2^2 in the case of an observed frequency distribution or μ_4/σ^4 in the case of a probability distribution is known as the *kurtosis* of the distribution and can be regarded as a measure of its degree of 'peakedness'. We can imagine two symmetrical distributions with the same mean and variance, the first of which has long tails and rises to a high, narrow peak and the second of which has short tails and a low, broad peak. The first is called leptokurtic (from the Greek λεπτός = thin) and will have a large kurtosis because the large deviations in the tails will have a big effect when raised to the fourth power; the second is called platykurtic (from the Greek πλατύς = broad) and will have a small kurtosis.

The kurtosis must clearly be positive since fourth powers are always positive. An example of an extremely platykurtic distribution is provided by the uniform distribution whose fourth moment is

$$\mu_4 = \int_0^1 (x-\tfrac{1}{2})^4 dx = [\tfrac{1}{5}(x-\tfrac{1}{2})^5]_0^1 = \tfrac{1}{80}$$

and whose kurtosis is therefore

$$\frac{\mu_4}{\sigma^4} = \frac{144}{80} = 1{\cdot}8.$$

The smallest kurtosis possible is 1; this value is attained by a random variable which takes only two values with equal probabilities (see Problem 4.2). At the other end of the scale a highly leptokurtic distribution may have an infinite fourth moment, and consequently an infinite kurtosis (provided its variance is finite); the t distribution with four degrees of freedom is an example of such a distribution. The normal distribution has a kurtosis of 3 and this is consequently taken as a standard of comparison; distributions with a kurtosis greater than 3 are described as leptokurtic and distributions with a kurtosis less than 3 as platykurtic. A useful *aide-memoire* due to 'Student' (W. S. Gosset) is shown in Fig. 15. The kurtosis is not of much use for comparing the shapes of distributions of different degrees of symmetry.

FIG. 15. Platykurtic curves have short tails like a platypus, while leptokurtic curves have long tails like kangaroos noted for 'lepping' (after 'Student')

Exercises

4.1. Find the mean and the median of the 50 observations obtained in Exercise 3.2 and compare them with their theoretical values.

4.2. Find the mean and the median of the distributions of (*a*) head breadth (*b*) head length in Table 10 on p. 40. [Each measurement was taken to the nearest millimetre.]

4.3. Calculate the variance of the distribution of litter size in Table 8 on p. 29 (*a*) from the formula

$$S^2 = \Sigma x^2 n(x) - n\bar{x}^2,$$

and (*b*) by taking a 'working mean' of 6, that is to say by taking 6 from each litter size before squaring it and also from the mean before squaring it. [This simplifies the arithmetic and makes no difference to the answer because the variance is unaffected by a change of origin.]

4.4. If T is the sum of the observations, show that $n\bar{x}^2 = T^2/n$. It is often convenient to use this alternative formula for the correction factor to the sum of squares since the introduction of decimals is left to the end. Re-work the preceding exercise in this way.

4.5. Find the variance of the distribution of coin weights in Table 9 on p. 34.

4.6. In the distribution of head breadths in Table 10 on p. 40 code the class 13– as -3, $13\frac{1}{2}$– as -2 and so on and work out the mean and variance of this coded distribution. To find the real mean, the coded mean must be divided by 2 (since the class interval is only $\frac{1}{2}$ centimetre) and then be added to 14·7, the centre of the zero coded class; check that this gives the same answer as in Exercise 4.2(a). To find the real variance, divide the coded variance by 4, since the variance is in squared units. [Note that this method of coding could not be used for the distribution of coin weights in Table 9 since the centres of the class intervals are not evenly spaced.]

4.7. Find the mean and the variance of the distribution of head lengths in Table 10 (p. 40) by suitable coding.

4.8. Find the mean deviation, the interquartile range and the standard deviation of the 50 observations obtained in Exercise 3.2 and compare them with their theoretical values.

4.9. Find the interquartile range of the distribution of head length in Table 10 on p. 40, divide it by the standard deviation and compare this ratio with that of 1·349 for a normal distribution.

4.10. Find the skewness and kurtosis of the distribution of litter size in Table 8 on p. 29, using a working mean of 6.

4.11. Find the skewness and kurtosis of the probability distribution of the number of points on two dice.

4.12. In calculating the mean and variance of a grouped frequency distribution it is assumed that all the observations in a class interval are concentrated at the middle of that interval. This assumption has no systematic effect on the mean, but it tends to lead to a slight over-estimate of the variance, particularly if the grouping is coarse. To counteract this effect Sheppard suggested that the quantity $c^2/12$, where c is the unit of grouping, should be subtracted from the observed variance. Calculate the variance, with Sheppard's correction, for the distributions of (a) head breadth and (b) head length considered in Exercise 4.6 and 4.7.

Problems

4.1. Schwarz's inequality states that, for arbitrary functions $g(x)$ and $h(x)$,

$$\left[\int g(x)h(x)dx\right]^2 \leqslant \int g^2(x)dx \int h^2(x)dx.$$

Prove this inequality by observing that

$$\int [tg(x)-h(x)]^2 dx \geqslant 0 \text{ for all t,}$$

and that if a quadratic function in t, $\phi(t) = at^2+bt+c$, is positive for all values of t the equation $\phi(t) = 0$ can have no real solution and so b^2-4ac must be negative. What is the condition for the equality to hold?

4.2. Consider a random variable with density function $f(x)$ and with mean μ. By appropriate choice of $g(x)$ and $h(x)$ in Schwarz's inequality show that (*a*) the kurtosis cannot be less than 1, (*b*) the kurtosis cannot be less than the square of the skewness.

4.3. Suppose that a distribution is formed from the mixture in equal parts of two underlying distributions which are both normal with the same variance σ^2 but with different means, μ_1 and μ_2. Find the mean, variance, skewness and kurtosis of the mixed distribution and the condition for it to be bimodal. (See Chapter 7 for the definition and properties of the normal distribution.)

4.4. Show that the mean deviation of a set of measurements is a minimum when the deviations are measured from the median, but that the standard deviation is a minimum when they are measured from the mean.

4.5. If $f(x)$ and $g(x)$ are positive functions of x prove that

$$\int_{-\infty}^{\infty} g(x)f(x)dx \geqslant \int_{g(x)\geqslant a} g(x)f(x)dx \geqslant a\int_{g(x)\geqslant a} f(x)dx$$

and hence derive Tchebychef's inequality

$$\text{Prob}\,[|X-\mu| \geqslant k\sigma] \leqslant \frac{1}{k^2}.$$

Thus the probability that a random variable will deviate from the mean by more than 2σ is less than $\frac{1}{4}$, the probability that it will deviate by more than 3σ is less than $\frac{1}{9}$, and so on. This inequality is usually very conservative, but it is useful because it is true for *any* distribution.

CHAPTER 5

EXPECTED VALUES

It was stated in the last chapter that one of the main reasons for using the mean and the variance as measures of the location and dispersion of a probability distribution is their mathematical tractability. This is due to the fact that they can both be regarded as examples of *Expected values*, whose properties will be discussed in this chapter.

EXPECTED VALUES

The mean of a probability distribution is often called the *Expected value* of the random variable, denoted by $E(X)$, since it is the value of the average, $\bar{x}$, which one would expect to find in a very large sample. The formula for the Expected value is therefore

$$E(X) = \sum_x xP(x) \text{ for a discrete variable}$$

or
$$E(X) = \int xf(x)dx \text{ for a continuous variable.}$$

If $Y = \phi(X)$ is some function of X, such as X^2 or $\log X$, then the Expected value of Y is

$$E(Y) = E[\phi(X)] = \sum_x \phi(x)P(x) \text{ for discrete variable.}$$

$$E(Y) = E[\phi(X)] = \int \phi(x)f(x)dx \text{ for a continuous variable.}$$

This is the value which one would expect to find if one took a very large sample of observations, $x_1, x_2, \ldots, x_n$, calculated $y_i = \phi(x_i)$ for each observation and then computed the mean of the y_i's. The variance of the distribution is the Expected value of $(X-\mu)^2$, the third moment about the mean is the Expected value of $(X-\mu)^3$, and so on. We shall now consider the mathematical properties of Expected values. It will be assumed that the random variable is discrete; to prove the following theorems for continuous variables it is only necessary to replace $P(x)$ by $f(x)dx$ and the summation sign by an integral sign.

Firstly, the Expected value is invariant under any linear transformation of the random variable; that is to say, if a and b are any constants and if $Y = a+bX$, then

$$E(Y) \equiv E(a+bX) = a+bE(X).$$

This is intuitively obvious since, if we multiply all the observations by a constant, b, their mean will also be multiplied by b; and if we add a constant, a, to all the observations, their mean will also be increased by this amount. To prove this theorem rigorously we observe that, if X is a discrete random variable, then

$$\begin{aligned} E(a+bX) &= \sum(a+bx)P(x) = \sum aP(x)+\sum bxP(x) \\ &= a\sum P(x)+b\sum xP(x) = a+bE(X). \end{aligned}$$

It is important to remember that Expected values are not invariant under non-linear transformations; for example, the Expected value of X^2 is *not* equal to the square of the Expected value of X $[E(X^2) \neq E^2(X)]$. This point was discussed in the last chapter when the properties of means and medians were compared.

It follows from this theorem that the variance of $a+bX$ is b^2 times the variance of X; for, if we write $V(X)$ for the variance of X and μ for its Expected value, then

$$\begin{aligned} V(a+bX) &\equiv E[(a+bX-a-b\mu)^2] \\ &= E[b^2(X-\mu)^2] = b^2E[(X-\mu)^2] = b^2V(X). \end{aligned}$$

This result is also intuitively obvious since, if we multiply all the observations by b, then their variance, which is in square units, will be multiplied by b^2; and if we add a constant, a, to all the observations we shall leave the deviations from the mean unchanged. For example, if the variance of a distribution of heights, measured in inches, is 6 square inches and if we multiply all the observations by 2·54 to convert them to centimetres, the variance will become $2{\cdot}54^2 \times 6 = 38{\cdot}7$ square centimetres; if, however, we subtracted 12 inches from all the observations we should also subtract 12 inches from the mean so that the variance would be unaltered.

The second important theorem is that, if X and Y are any two random variables, then the Expected value of their sum is equal to the sum of their Expected values:

$$E(X+Y) = E(X)+E(Y).$$

This theorem is also intuitively obvious. For suppose that n paired observations (x_1, y_1), (x_2, y_2), ..., (x_n, y_n) have been made, and that for each pair, (x_i, y_i), their sum $w_i = x_i+y_i$ is calculated; it is clear that

$$\sum_{i=1}^{n} w_i \equiv \sum_{i=1}^{n} (x_i+y_i) = \sum_{i=1}^{n} x_i+ \sum_{i=1}^{n} y_i$$

from which it follows that $\bar{w} = \bar{x}+\bar{y}$. The same relationship must hold between the limiting values of these quantities in very large samples. For example, suppose that X and Y are the sitting height and leg length of a man, so that their sum is his total height; it is clear that the average height of a group of men is equal to their average sitting height plus their average leg length. To prove this theorem rigorously let us suppose that X and Y are discrete variables with joint probability function $P(x, y)$ and with marginal probability functions $P_1(x)$ and $P_2(y)$ respectively; then

$$\begin{aligned} E(X+Y) &= \sum_{x,y} (x+y)P(x,y) = \sum_x\sum_y xP(x,y) + \sum_y\sum_x yP(x,y) \\ &= \sum_x x\sum_y P(x,y) + \sum_y y\sum_x P(x,y) = \sum_x xP_1(x) + \sum_y yP_2(y) \\ &= E(X)+E(Y). \end{aligned}$$

Clearly this theorem can be extended to any number of random variables. It is important because one often wishes to study the properties of sums of random variables, particularly in the theory of sampling distributions, which are fundamental in statistical inference. It also enables us to prove some useful relationships between the moments about the mean and the moments about the origin of a probability distribution.

The rth moment about the origin, μ'_r, is the Expected value of X^r

$$\mu'_r = E(X^r).$$

The rth moment about the mean, μ_r, is the Expected value of the rth power of the deviation of X from its mean value, μ:

$$\mu_r = E[(X-\mu)^r].$$

Thus μ'_1, is the same as the mean, μ; μ_2 is the variance and μ_3 and μ_4 are the third and fourth moments about the mean used in constructing measures of skewness and kurtosis. It is often easier to compute the moments about the origin and therefore it is useful to have formulae which relate the two kinds of moments. Consider first the variance:

$$\mu_2 = E[(X-\mu)^2] = E(X^2-2\mu X+\mu^2)$$
$$= E(X^2)-2\mu E(X)+\mu^2 = \mu'_2-\mu^2.$$

Thus for the uniform distribution, $\mu = \frac{1}{2}$ and

$$\mu'_2 = \int_0^1 x^2dx = [\tfrac{1}{3}x^3]_0^1 = \tfrac{1}{3}$$

from which it follows that $\mu_2 = \frac{1}{3}-\frac{1}{4} = \frac{1}{12}$. The same result has already been obtained by direct calculation. It can be shown in a similar way that

$$\mu_3 = \mu'_3-3\mu'_2\mu+2\mu^3$$

and

$$\mu_4 = \mu'_4-4\mu'_3\mu+6\mu'_2\mu^2-3\mu^4.$$

These results are, of course, analogues of the results already derived for the observed moments of a frequency distribution.

The third important theorem about Expected values is that if X and Y are independently distributed then the Expected value of their product is equal to the product of their Expected values:

$$E(XY) = E(X)\cdot E(Y).$$

It is important to remember that this result will not usually be true unless the variables are independent. To prove this theorem let us suppose that X and Y are independent discrete random variables; then

$$E(XY) = \sum_{x,y} xyP(x,y) = \sum_x\sum_y xyP_1(x)P_2(y)$$
$$= \sum_x xP_1(x)\sum_y yP_2(y) = E(X)\cdot E(Y).$$

An important consequence of this theorem is that if X and Y are independently distributed, then the variance of their sum is equal to the sum of their variances:

$$V(X+Y) = V(X)+V(Y).$$

For suppose that the Expected values of X and Y are ξ and η respectively; then

$$\begin{aligned} V(X+Y) &= E[(X+Y-\xi-\eta)^2] \\ &= E[((X-\xi)+(Y-\eta))^2] \\ &= E[(X-\xi)^2+(Y-\eta)^2+2(X-\xi)(Y-\eta)] \\ &= E(X-\xi)^2+E(Y-\eta)^2+2E(X-\xi) \,.\, E(Y-\eta) \\ &= V(X)+V(Y). \end{aligned}$$

For example, if X and Y are the numbers of points on two unbiased dice, then $X+Y$ is the sum of the number of points on both dice, whose distribution is shown in Table 4 on p. 20. Now $E(X) = 3\frac{1}{2}$ and

$$E(X^2) = \tfrac{1}{6}(1^2+2^2+3^2+4^2+5^2+6^2) = 15\tfrac{1}{6}$$

so that

$$V(X) = E(X^2)-E^2(X) = 15\tfrac{1}{6}-(3\tfrac{1}{2})^2 = \tfrac{35}{12}.$$

Hence

$$V(X+Y) = V(X)+V(Y) = \tfrac{35}{12}+\tfrac{35}{12} = 5\tfrac{5}{6}.$$

The same result has already been found by direct computation.

The variance of the difference between two independent random variables, $X-Y$, is also equal to the sum of their variances since $V(-Y) = V(Y)$. This result can obviously be extended to any number of independent random variables.

In this section we have proved three important theorems about Expected values:

(1) If X is any random variable and if a and b are constants,

$$E(a+bX) = a+bE(X).$$

(2) If X and Y are any pair of random variables,

$$E(X+Y) = E(X)+E(Y).$$

(3) If X and Y are any pair of independent random variables,

$$E(XY) = E(X) \cdot E(Y).$$

It follows from the first theorem that

$$V(a+bX) = b^2 V(X)$$

and from the third theorem that

$$V(X+Y) = V(X)+V(Y)$$

if X and Y are independent. These results enable us to find the mean and the variance of the sum, or of any linear function, of any number of independent random variables.

Covariance and Correlation

The covariance of a pair of random variables is defined as the Expected value of the product of their deviations from their means:

$$\text{Cov}\ (X, Y) = E[(X-\xi)(Y-\eta)].$$

If the variables are independent the covariance is zero since

$$E[(X-\xi)(Y-\eta)] = E(X-\xi)\ .\ E(Y-\eta) = 0.$$

This fact has already been used in proving that the variance of the sum of two independent random variables is equal to the sum of their variances. However, if the variables are not independent the covariance will in general not be zero and the formula for the variance of $X+Y$ must be re-written as

$$V(X+Y) = V(X)+V(Y)+2\ \text{Cov}\ (X, Y).$$

As its name implies the covariance measures the extent to which X and Y vary together. For example, it is clear from Table 10 on p. 40 that men with long heads tend to have broader heads than men with short heads. The covariance of head length and head breadth will therefore be positive since a positive deviation of head length from its mean will more often than not be accompanied by a positive deviation of head breadth; likewise negative deviations of the two variables will tend to go together, which when multiplied together produce again a positive value. If, on the other hand, large values of one variable tended to be accompanied by small values of the other, and *vice versa*, the covariance would be negative. It is possible, though unlikely in practice, for dependent variables to have a zero covariance (see Exercise 5.4).

The theoretical covariance is the limiting value of the observed covariance of a frequency distribution defined as

$$\text{covariance} = \frac{\sum_{i=1}^{n} (x_i - \bar{x})(y_i - \bar{y})}{n}$$

if the individual paired observations are available, or by the analogous formula for a grouped bivariate frequency distribution. In practice the covariance is most often calculated in order to find the correlation coefficient, r, which is defined as the covariance divided by the product of the standard deviations of the two distributions:

$$r = \frac{\Sigma(x_i - \bar{x})(y_i - \bar{y})/n}{\sqrt{\frac{\Sigma(x_i - \bar{x})^2}{n} \frac{\Sigma(y_i - \bar{y})^2}{n}}} = \frac{\Sigma(x_i - \bar{x})(y_i - \bar{y})}{\sqrt{\Sigma(x_i - \bar{x})^2 \Sigma(y_i - \bar{y})^2}}.$$

As the number of observations increases the observed correlation coefficient, r, tends towards its theoretical counterpart, ρ, defined as

$$\rho = \frac{\text{Cov}(X, Y)}{\sqrt{V(X) \cdot V(Y)}}.$$

The correlation coefficient is a dimensionless quantity which can therefore be used, with certain reservations, as an absolute measure of the relationship between the two variables. If X and Y are independent then their covariance and their correlation coefficient are zero. If large values of X tend to be accompanied by large values of Y and *vice versa*, then their covariance and their correlation coefficient will be positive; in the extreme case when one variable is completely determined by the other one in such a way that the points (x_i, y_i) lie exactly on a straight line with a positive slope, the correlation coefficient attains its largest possible value of $+1$. (See

Exercise 5.6 and Problem 5.3.) On the other hand, if large values of X tend to be accompanied by small values of Y and *vice versa*, their covariance and correlation coefficient will be negative; in the extreme case when the points (x_i, y_i) lie exactly on a straight line with a negative slope, the correlation coefficient attains its smallest possible value of -1. Provided that the relationship between the two variables is approximately linear the correlation coefficient provides a reasonable measure of the degree of association between them, but may underestimate it if their relationship is non-linear (see Exercise 5.4). The interpretation of correlation coefficients will be discussed further in Chapter 12.

The Moment Generating Function

We shall now consider a device which is often very useful in the study of probability distributions. The moment generating function (m.g.f.) of a random variable X is defined as the Expected value of e^{tX},

$$M(t) = E(e^{tX}).$$

The m.g.f., $M(t)$, is thus a function of the auxiliary mathematical variable t. The reason for defining this function is that it can be used to find all the moments of the distribution. For the series expansion of e^{tX} is

$$e^{tX} = 1+tX+\frac{t^2X^2}{2!}+\frac{t^3X^3}{3!}+\ldots.$$

Hence

$$M(t) = E(e^{tX}) = 1+t\mu'_1+\frac{t^2\mu'_2}{2!}+\frac{t^3\mu'_3}{3!}+\ldots.$$

If we differentiate $M(t)$ r times with respect to t and then set $t = 0$ we shall therefore obtain the rth moment about the origin μ'_r.

For example, the m.g.f. of the uniform distribution is

$$M(t) = E(e^{tX}) = \int_0^1 e^{tx}dx = \left[\frac{1}{t}e^{tx}\right]_0^1 = \frac{1}{t}\left[e^t-1\right]$$

$$= 1+\frac{t}{2!}+\frac{t^2}{3!}+\frac{t^3}{4!}+\ldots.$$

If we differentiate this series r times and then set $t = 0$ we find that $\mu'_r = \frac{1}{r+1}$. In particular $\mu'_1 = \mu = \frac{1}{2}$ and $\mu'_2 = \frac{1}{3}$, as we have already found by direct calculation.

The moment generating function thus provides a powerful general method of determining the moments of a probability distribution. Its importance in statistical theory does not end here for it also enables us to find the distribution of the sum of two or more independent random variables. Suppose that X and Y are independent random variables with moment generating functions $M_1(t)$ and $M_2(t)$ respectively. The m.g.f. of their sum, $X+Y$, is

$$M(t) = E[e^{t(X+Y)}] = E(e^{tX}e^{tY}) = E(e^{tX})\cdot E(e^{tY})$$
$$= M_1(t)\cdot M_2(t).$$

Thus the m.g.f. of the sum of two independent random variables is equal to the product of their m.g.f.'s. This important theorem can obviously be extended to any number of independent random variables.

For example, if X and Y are independent random variables following the uniform distribution the m.g.f. of their sum, $X+Y$, is $\frac{1}{t^2}(e^t-1)^2$. This is also the m.g.f. of a random variable which follows the 'triangular' distribution whose density function is an isosceles triangle:

$$f(x) = x \qquad 0 \leqq x \leqq 1$$
$$f(x) = 2-x \qquad 1 \leqq x \leqq 2$$

[The m.g.f. of this random variable can be obtained quite easily by using the result that the indefinite integral of xe^{tx} is $\frac{1}{t^2}(xt-1)e^{tx}$.] It follows that $X+Y$ must follow this distribution, since it is scarcely conceivable that two different distributions could have the same m.g.f. (See Problem 3.5 for a direct proof of this result.) This method is very useful in proving results about the distribution of sums of random variables, although it is necessary either to know the answer beforehand or to be able to recognise the distribution to which the moment generating function belongs.

Another useful result is that if a random variable X has m.g.f. $M(t)$ then the m.g.f. of any linear function of X, say $Y = a+bX$, is $e^{at}M(bt)$; for

$$E(e^{tY}) = E(e^{(a+bX)t}) = E(e^{at}e^{btX}) = e^{at}E(e^{btX}) = e^{at}M(bt).$$

In particular, if $\mu = E(X)$ then $e^{-\mu t}M(t)$ is the m.g.f. of $X-\mu$ which will generate the moments about the mean in the same way that $M(t)$ generates the moments about the origin.

These two theorems about the m.g.f. of a linear function of a random variable and about the m.g.f. of the sum of two independent random variables enable us to find the m.g.f. of any linear function of any number of independent random variables; if the latter can be recognised as the m.g.f. of a known distribution we have solved the problem of finding the distribution of the linear function.

It must be remembered that some random variables do not possess a moment generating function; for example, neither the t distribution nor the F distribution possesses a moment generating function since some of their moments are infinite. To get round this difficulty mathematical statisticians use the characteristic function defined as

$$\phi(t) = E(e^{itX}).$$

This function always exists since $|e^{itx}| \leq 1$ but it is beyond the scope of this book since it employs the ideas of complex numbers and complex integrals.

Exercises

5.1. Evaluate the mean and the variance of the distribution of the number of points on 3 dice (*a*) directly (see Exercise 2.3), (*b*) by using the mean and variance of the distribution of the number of points on a single die (see p. 72).

5.2. If the silver pennies in Table 9 on p. 34 were weighed in randomly selected groups of five, what would be the mean and standard deviation of the resulting distribution? [The mean and variance of the original distribution are given on p. 46 and in Exercise 4.5 respectively.]

5·3. Prove that $\text{Cov}(X, -Y) = -\text{Cov}(X, Y)$ and hence show that

$$V(X-Y) = V(X)+V(Y) - 2\,\text{Cov}(X, Y).$$

If the distribution of men's height has a mean of 5 ft 8 in and a standard deviation of $2\frac{1}{2}$ in and if the heights of brothers have a correlation of $\frac{1}{2}$, what is the mean and variance of the difference in height between pairs of brothers? of the sum of their heights?

5·4. Suppose that X is a random variable taking the values -1, 0 and 1 with equal probabilities and that $Y = X^2$. Show that X and Y are uncorrelated but not independent.

5·5. Prove that

$$\Sigma(x_i - \bar{x})(\bar{y}_i - \bar{y}) = \Sigma x_i \bar{y}_i - n\bar{x}\bar{y}.$$

Use this fact to find, by suitable coding, the covariance and the correlation coefficient between head length and head breadth from the data in Table 10 on p. 40 (see Exercises 4.6 and 4.7.)

5·6. If the paired observations $(x_1, y_1), (x_2, y_2), \ldots, (x_n, y_n)$ lie exactly on the straight line $y_i = a+bx_i$, show that $(y_i - \bar{y}) = b(x_i - \bar{x})$ and hence show that the correlation coefficient is $+1$ or -1 according as b is positive or negative.

Problems

5·1. If X and Y are independent random variables, find the third and fourth moments about the mean of $X+Y$ in terms of the moments of X and Y.

5·2. If two quantities X and Y have a correlation ρ and can each be regarded as being the sum of k independent components with the same distribution, h of which are common to X and Y, what is the relation between ρ, k and h? If, further, it is required to predict the value of Y from the value of X, what is the mean square prediction error, $E(Y-X)^2$, in terms of ρ and the variance of Y? [Certificate, 1958]

5·3. (*a*) Observe that $\Sigma[(x_i - \bar{x}) + t(y_i - \bar{y})]^2 \geqslant 0$ for all t and hence show that $r^2 \leqslant 1$. [Cf. Problem 4.1; this is equivalent to proving Cauchy's inequality, $(\Sigma a_i b_i)^2 \leqslant \Sigma a_i^2 \Sigma b_i^2$.] (*b*) Observe that $V(X+ty) \geqslant 0$ for all t and hence show that $\rho^2 \leqslant 1$. (Cf. Problem 4.1.)

5·4. If X is a random variable with Expected value ξ and if Y is some function of X, $Y = \phi(X)$, it is not in general true that $E(Y) = \phi(\xi)$. However, if we write $X = \xi + \varepsilon$, then to a first approximation $\phi(X) = \phi(\xi) + \varepsilon\phi'(\xi)$. Hence show that

$$E(Y) \doteqdot \phi(\xi)$$

$$V(Y) \doteqdot V(X)\phi'^2(\xi).$$

This method of finding the approximate mean and variance of a function of a random variable is called the *Delta technique*.

5.5. The time taken for a complete swing of a pendulum of length l is $2\pi\sqrt{(l/g)}$, where g is an unknown constant. A physicist takes 20 independent measurements, each measurement being of the time taken by a 2-ft pendulum to complete 100 consecutive swings and he finds that the mean and the standard deviation of these 20 measurements are 157·1 and 0·4 seconds respectively. Estimate g and attach a standard error to your estimate. [Certificate, 1954]

5.6. Suppose that X and Y are random variables with Expected values ξ and η. Write $X = \xi + \varepsilon_1$, $Y = \eta + \varepsilon_2$, and, by ignoring error terms of degree higher than the first, show that

$$E(XY) \doteqdot \xi\eta$$

$$V(XY) \doteqdot \xi^2\eta^2 \left\{ \frac{V(X)}{\xi^2} + \frac{V(Y)}{\eta^2} + \frac{2\mathrm{Cov}(X, Y)}{\xi\eta} \right\}.$$

5.7. Show similarly that

$$E\left(\frac{X}{Y}\right) \doteqdot \frac{\xi}{\eta}$$

$$V\left(\frac{X}{Y}\right) \doteqdot \frac{\xi^2}{\eta^2} \left\{ \frac{V(X)}{\xi^2} + \frac{V(Y)}{\eta^2} - \frac{2\mathrm{Cov}(X, Y)}{\xi\eta} \right\}.$$

(First expand $\frac{1}{Y}$ about $Y = \eta$.)

5.8. The velocity of sound in a gas is given by the formula

$$u = \sqrt{\left(\frac{ps_1}{ds_2}\right)},$$

when p is the pressure, d the density and s_1 and s_2 are the two specific heats. Supposing p, d, s_1 and s_2 may have small errors of measurement, each distributed independently with zero mean and small coefficient of variation, find the coefficient of variation of u. [Certificate, 1957]. (The coefficient of variation is the ratio of the standard deviation to the mean.)

5.9. For a discrete random variable taking the integral values 0, 1, 2, ... it is often better to use the probability generating function (p.g.f.) rather than the moment generating function. The p.g.f. is defined as

$$G(s) = E(s^X) = P(0) + P(1)s + P(2)s^2 + \dots$$

Prove that

(*i*) $G^{(r)}(0) = r!\,P(r)$. Thus if the p.g.f. is known the whole probability distribution can be calculated.

(*ii*) $G^{(r)}(1) = E[X(X-1)(X-2)\ldots(X-r+1)]$. This is the rth factorial moment, from which the ordinary moments can be calculated.

(*iii*) If X and Y are independent integral-valued random variables, then the p.g.f. of $X+Y$ is equal to the product of their p.g.f.'s.

Chapter 6

THE BINOMIAL, POISSON AND EXPONENTIAL DISTRIBUTIONS

In this chapter we shall consider three important probability distributions, the binomial, Poisson and exponential distributions. The first two have long histories in statistical theory, the binomial distribution having been discovered by James Bernoulli about 1700 and the Poisson distribution, which is a limiting form of the binomial distribution, by S. D. Poisson in 1837. The exponential distribution, which can be regarded as the continuous analogue of the Poisson distribution, has achieved importance more recently in connection with the theory of stochastic processes. The normal distribution, which is probably the most important distribution in statistical theory, will be considered separately in Chapter 7.

The Binomial Distribution

Imagine an experiment which can be repeated indefinitely and which at each repetition, or trial, results in one of two possible outcomes. For example, the experiment might consist in tossing a coin, in which case the two outcomes would be heads and tails, or in observing the sex of a baby, or in throwing a die and recording whether or not the throw is a six; in general the two possible outcomes will be called *success* (S) and *failure* (F). We shall suppose that at each trial there is a certain probability, P, of success, and a corresponding probability, $Q = 1 - P$, of failure; and we shall suppose furthermore that the probability of success remains constant from trial to trial and that the results of successive trials are statistically independent of each other. Imagine now that the experiment has been repeated n times, resulting in a series of successes and failures, $SSFSFF$. . . . In many situations the order in which the successes and failures occur is irrelevant and the experimenter is only interested in the total number of successes. It is therefore natural to ask: What is the probability of obtaining x successes and $n-x$ failures in n repetitions of the experiment?

To fix our ideas let us return to the coin-tossing experiment of Kerrich described on p. 2. The original sequence of 10,000 spins of a coin can be regarded as built up of 2000 sequences each consisting of five spins. Now in five spins of a coin there can be 0, 1, 2, 3, 4 or 5 heads, and the number of heads is thus a random variable which can take one of these six values. What should be the probability distribution of this random variable on the assumptions that at each spin the probability of heads is $\frac{1}{2}$, and that the results of successive spins are statistically independent of each other?

The chance of throwing no heads in 5 spins is the chance of throwing five tails running which is $(\frac{1}{2})^5 = 1/32$. The chance of throwing one head and four tails *in a particular order*, such as $THTTT$ is likewise $(\frac{1}{2})^5 = 1/32$; but there are five possible positions in which the single head may occur and so the overall probability of obtaining 1 head in 5 spins is 5/32. Similarly the probability of obtaining 2 heads and 3 tails is 10/32 since this can happen in any of the following ways, each of which has a probability of 1/32:

$HHTTT$	$HTHTT$	$HTTHT$	$HTTTH$
	$THHTT$	$THTHT$	$THTTH$
		$TTHHT$	$TTHTH$
			$TTTHH$

The probabilities of 3, 4 and 5 heads can be found in a similar way, and the probabilities of obtaining 0, 1, 2, 3, 4 and 5 heads are thus 1/32, 5/32, 10/32, 10/32, 5/32 and 1/32 respectively. This probability distribution is compared in Table 11

TABLE 11

The distribution of the number of heads in 2000 sequences of 5 spins of a coin

No. of heads	0	1	2	3	4	5	Total
Frequency	59	316	596	633	320	76	2000
Relative frequency	·030	·158	·298	·316	·160	·038	1·000
Theoretical probability	·031	·156	·312	·312	·156	·031	1·000

with the observed frequency distribution from Kerrich's experiment; it will be seen that there is good agreement between the two.

We will now consider the general case of n trials at each of which there is a probability, P, of 'success' and a probability $Q = 1-P$, of 'failure'. The chance of obtaining x successes and $n-x$ failures in a *particular order*, such as x successes followed by $n-x$ failures, is, by the Law of Multiplication, $P^x Q^{n-x}$. To find the overall probability of obtaining x successes, in any order, we must multiply this quantity by the number of ways in which this event can happen. In any particular case we could evaluate the number of different ways of obtaining x successes and $n-x$ failures by direct enumeration, but this procedure would clearly become rather tedious and it will be convenient to find a general formula. This requires a brief digression on the theory of combinations.

n distinct objects can be rearranged among themselves in $n(n-1)(n-2)...3\times2\times1$ different ways. For the first object can be chosen in n ways; once the first object has been chosen, the second object can be chosen in $n-1$ ways; then the third object can be chosen in $n-2$ ways, and so on. This number is called 'factorial n' and is usually written $n!$ for short. For example, the letters A, B, C can be rearranged in $3! = 3\times2\times1 = 6$ ways, which are

$$\begin{matrix} ABC & BAC & CAB \\ ACB & BCA & CBA \end{matrix}$$

If, however, the objects are not all distinct, the number of rearrangements is reduced; for example, there are only three ways of rearranging the letters A, B, B (ABB, BAB, BBA).

To find the number of ways of obtaining x successes in n trials let us write down x S's followed by $n-x$ F's and number them from 1 to n:

$$\begin{matrix} S & S & S & \ldots & S & F & F & \ldots & F \\ 1 & 2 & 3 & & x & x+1 & x+2 & & n \end{matrix}$$

Each rearrangement of the numbers 1 to n will define a corresponding rearrangement of the S's and F's; for example, the rearrangement (n, 2, 3, ..., $n-1$, 1) corresponds to *FSS...FFS*. However, not every arrangement of the numbers 1 to n leads to a different arrangement of the S's and F's; in fact all arrangements which merely rearrange the numbers 1 to x and the numbers $x+1$ to n among themselves correspond to the same arrangement of the S's and F's. Now there are $x!$ ways of rearranging the first x numbers and $(n-x)!$ ways of rearranging the last $(n-x)$ numbers among themselves, and so to each different arrangement of the S's and F's there correspond $x!(n-x)!$ rearrangements of the numbers 1 to n which give rise to it. Since there are altogether $n!$ rearrangements of the numbers 1 to n, it follows that there are

$$\frac{n!}{x!(n-x)!}$$

different ways of rearranging the S's and F's. To check that this formula gives the correct answer in a case which we have already considered by direct enumeration, we note that there are

$$\frac{5!}{1!4!} = \frac{120}{24} = 5$$

ways of obtaining 1 success and 4 failures in 5 trials, and

$$\frac{5!}{2!3!} = \frac{120}{2\times 6} = 10$$

ways of obtaining 2 successes and 3 failures. To obtain the correct answer of 1 when $x = 0$ or $x = n$ we must define $0!$ as 1.

We have now found the number of ways in which x successes can occur in n trials, and we have seen that the probability of each of them is P^xQ^{n-x}; it follows by the Law of Addition that the probability of x successes in n trials is

$$P(x) = \frac{n!}{x!(n-x)!} P^xQ^{n-x}$$
$$x = 0, 1, 2, ..., n.$$

This probability distribution is called the *binomial distribution* because the probabilities $P(0)$, $P(1)$, and so on, are the successive terms in the expansion of $(Q+P)^n$ by the binomial theorem:

$$(Q+P)^n = Q^n+nQ^{n-1}P+\ldots+nQP^{n-1}+P^n$$
$$= P(0)+P(1)+\ldots+P(n-1)+P(N).$$

This shows immediately that the probabilities add up to 1, as they should do, since $Q+P = 1$. (If the n brackets $(Q+P)(Q+P)\ldots(Q+P)$ are multiplied out the coefficient of Q^xP^{n-x} must be the number of ways of choosing Q from x of the brackets and P from the remainder.)

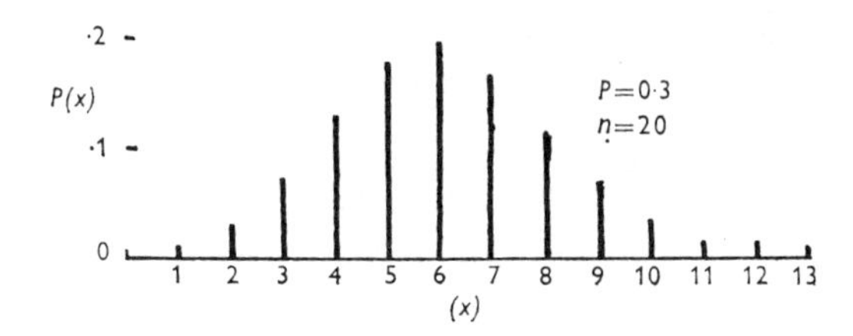

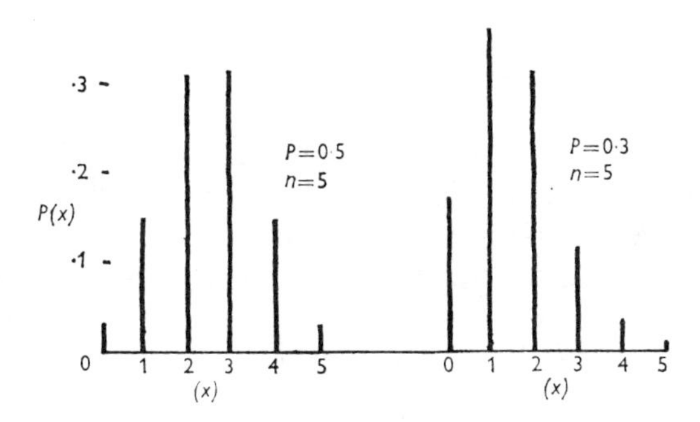

FIG. 16. The binomial distribution

Some examples of the binomial distribution with different values of P and n are shown in Fig. 16. It will be seen that, as one might expect, the distribution rises to a maximum value and then falls away again. It can be verified quite easily that

$$\frac{P(x-1)}{P(x)} = \frac{xQ}{(n-x+1)P}.$$

This quantity is less than 1 if xQ is less than $(n-x+1)P$, that is to say if x is less than $(n+1)P$, and greater than 1 otherwise. This means that $P(x)$ increases as x increases until x reaches the integral value immediately below $(n+1)P$ and then decreases. For example, if $n = 10$ and $P = \frac{1}{2}$, then $(n+1)P = 5\frac{1}{2}$ and so $P(x)$ increases until it attains its maximum value at $x = 5$ and then decreases. If $(n+1)P$ happens to be an integer the distribution will have two maximum values, at $(n+1)P-1$ and $(n+1)P$; for example, if $n = 5$ and $P = \frac{1}{2}$, then $(n+1)P = 3$ and so the distribution will have maximum values at 2 and 3, which can be verified from Table 11 on p. 82.

The moments of the binomial distribution are most easily found from the moment generating function which is

$$M(t) = E(e^{tX}) = \sum_{x=0}^{n} e^{tx} \frac{n!}{x!(n-x)!} P^x Q^{n-x}$$
$$= \sum_{x=0}^{n} \frac{n!}{x!(n-x)!} (Pe^t)^x Q^{n-x} = (Q+Pe^t)^n.$$

We find by differentiation that

$$M'(t) = nPe^t(Q+Pe^t)^{n-1}$$
$$E(X) = M'(0) = nP(Q+P)^{n-1} = nP.$$

It is in fact intuitively obvious that this is the mean of the distribution since if there are, on the average, P successes per trial, the average number of successes in n trials must be nP. By differentiating the m.g.f. again and setting $t = 0$ it will be found that

$$E(X^2) = M''(0) = nP+n(n-1)P^2$$

so that

$$V(X) = E(X^2)-E^2(X) = nP-nP^2 = nPQ.$$

This result can be proved rather more simply by using the fact that the variance of a sum of independent random variables is equal to the sum of their variances. The number of successes in n trials is clearly equal to the number of successes in the first trial plus the number of successes in the second trial and so on; that is to say

$$X = Z_1+Z_2+\dots+Z_n = \sum_{i=1}^{n} Z_i$$

where Z_i is the number of successes in the ith trial. Now Z_i is a random variable which takes the value 0 and 1 with probabilities Q and P respectively; hence

$$\begin{aligned} E(Z_i) &= P \\ E(Z_i^2) &= P \\ V(Z_i) &= E(Z_i^2) - E^2(Z_i) = P - P^2 = PQ. \end{aligned}$$

It follows that

$$E(X) = \sum_{i=1}^{n} E(Z_i) = nP$$

and, since the trials are independent,

$$V(X) = \sum_{i=1}^{n} V(Z_i) = nPQ.$$

This approach can also be used to derive the m.g.f. of X. For it is obvious that the m.g.f. of each of the Z_i's is $Q+Pe^t$; the m.g.f. of their sum is the product of their m.g.f.'s which is $(Q+Pe^t)^n$.

The variance of the distribution therefore increases as n increases. If, however, instead of the number of successes, X, we consider the proportion of successes, $Y = X/n$, which will take the values 0, $1/n$, $2/n$, ..., 1, with the appropriate binomial probabilities, then

$$E(Y) = E(X)/n = P$$

and

$$V(Y) = V(X)/n^2 = PQ/n.$$

Thus, whereas the variability of the *number* of successes in n trials increases, the variability of the *proportion* of successes decreases as n increases. This fact provides the answer to a popular fallacy known as the ' Law of Averages '. According to this ' Law ', if an unbiased coin has turned up heads ten times in succession, then it is more likely than not to turn up tails next time; for, it is argued, the numbers of heads and tails must in the long run be equal and so the deficit of ten tails must be made up sooner or later. It is not however the numbers but the proportions of heads and tails which will be equal in the long run. For example, in Kerrich's experiment (p. 2), there were 44 heads and 56 tails after a hundred spins, and

5067 heads and 4933 tails after 10,000 spins; between these two points the difference between the numbers of heads and tails has increased tenfold, but the difference between their proportions has decreased tenfold. What will happen to the ten heads in a row is not that they will be counterbalanced by an excess of ten tails in the succeeding spins, but that their effect on the proportion of heads will become progressively smaller as the number of spins increases.

Further differentiation of the m.g.f. shows that the third and fourth moments of the binomial distribution are

$$\begin{aligned} \mu_3 &= nPQ(Q-P) \\ \mu_4 &= 3n^2P^2Q^2+nPQ(1-6PQ) \end{aligned}$$

so that its measures of skewness and kurtosis are

$$\text{Skewness} = \frac{Q-P}{\sqrt{nPQ}}.$$

$$\text{Kurtosis} = 3+\frac{(1-6PQ)}{nPQ}.$$

The distribution is symmetrical only when $P = Q = \frac{1}{2}$. It is skew to the right when P is less than $\frac{1}{2}$ and skew to the left when P is greater than $\frac{1}{2}$, but the degree of asymmetry decreases as n increases (see Fig. 16 on p. 85). The kurtosis tends to the standard value of 3 as n increases.

We have already considered an application of the binomial distribution in a coin-tossing experiment. For a less artificial example we consider some well-known data on the number of boys and girls in families of different sizes obtained from the birth registrations in Saxony in the period 1876-1885; the data for families of 8 children are shown in Table 12. If the probability of a male birth is constant and is the same in all families, then the number of boys in a family of fixed size should follow the binomial distribution since the situation is exactly the same as in the spinning of a possibly biased coin. If, however, the probability of a male birth varies for any reason, either from one family to another or within the same family due to such factors as parental age, then there should be more families in the extreme classes than the binomial formula predicts; to take a *reductio ad absurdum*, if there were only two types of father, those capable of begetting only girls and those capable of begetting only boys, then there would be only two types of family, all boys and all girls.

To test whether or not Geissler's data follow a binomial distribution the expected numbers in the third column of Table 12 were calculated by substituting ·5147, the observed proportion of boys, for P in the binomial probabilities shown in the last column and then multiplying this predicted proportion by 53,680 to give a predicted number of families. Agreement with the observed data is quite good, although there are rather too many families of all boys or all girls.

TABLE 12

The number of boys in families containing 8 children (Geissler, 1889)

No. of boys	No. of families: Observed	No. of families: Expected	Binomial probability
0	215	165	Q^8
1	1,485	1,402	$8PQ^7$
2	5,331	5,203	$28P^2Q^6$
3	10,649	11,035	$56P^3Q^5$
4	14,959	14,628	$70P^4Q^4$
5	11,929	12,410	$56P^5Q^3$
6	6,678	6,580	$28P^6Q^2$
7	2,092	1,994	$8P^7Q$
8	342	264	P^8
Total	53,680	53,680	1

There has been some controversy whether this slight departure from theory is a biological reality or whether it is due to deficiencies in the information on the birth registers; subsequent studies with new data have come to conflicting conclusions. In any case, the departure is very slight and it can be concluded that the assumptions made in deriving the binomial distribution are very nearly, if not exactly, fulfilled in sex-determination in man.

It is rather difficult to find frequency distributions which follow the binomial distribution with any accuracy since most large bodies of data are heterogeneous. For example, it is almost certain that data on the fertility of turkeys' eggs arranged in the same form as Table 12 would not follow a binomial distribution since some turkeys are more fertile than others. The great importance of the binomial distribution in modern statistical theory lies in its use as a sampling distribution. If we have observed that, out of 50 eggs laid by one turkey, 40 are fertile and 10 infertile, this can be regarded as a single observation from a binomial distribution with $n = 50$ and an unknown probability, P, that this bird will lay a fertile egg. The standard methods of statistical inference can now be used to draw conclusions about P. This argument will be understood better when the reader has studied the chapters on statistical inference, but it should be clear that this sort of situation arises commonly and that the binomial distribution will therefore often be used in statistical inference.

The Poisson Distribution

The Poisson distribution is the limiting form of the binomial distribution when there is a large number of trials but only a small probability of success at each of them. If we write $\mu = nP$ for the mean number of successes and substitute μ/n for P in the formula for the binomial distribution we find after a little rearrangement that the probability of x successes is

$$P(x) = \left[\frac{n}{n}\cdot\frac{(n-1)}{n}\cdot\frac{(n-2)}{n}\cdots\frac{(n-x+1)}{n}\right]\frac{\mu^x}{x!}\left[\left(1-\frac{\mu}{n}\right)^{n-x}\right].$$

If we let n tend to infinity while keeping x and μ fixed, the terms $\frac{(n-1)}{n}, \frac{(n-2)}{n}, \ldots, \frac{(n-x+1)}{n}$ will all tend to 1 and the expression in the last square bracket will tend to $e^{-\mu}$, since $\left(1-\frac{\mu}{n}\right)^n \rightarrow e^{-\mu}$ and $\left(1-\frac{\mu}{n}\right)^{-x} \rightarrow 1$; hence

$$P(x) \rightarrow \frac{e^{-\mu}\mu^x}{x!} \qquad x = 0, 1, 2, \ldots.$$

This limiting distribution is called the *Poisson distribution*. It will be observed that it depends on μ but not on n and P separately.

The probabilities add up to 1 as they should do since

$$\sum_{x=0}^{\infty} P(x) = e^{-\mu} \sum_{x=0}^{\infty} \frac{\mu^x}{x!} = e^{-\mu}e^{\mu} = 1.$$

The function of the factor $e^{-\mu}$, which does not vary with x, is thus to ensure that the probabilities add up to 1. The moments of the distribution can be found by substituting μ for nP in the formulae for the moments of the binomial distribution and then letting P tend to 0 and Q to 1. Both the mean and the variance are equal to μ; the equality of the mean and the variance is a useful diagnostic feature of the distribution. The skewness is $1/\sqrt{\mu}$, so that the distribution is always skew to the right, but the degree of skewness decreases as μ increases. The kurtosis is $3+\frac{1}{\mu}$ which tends to the standard value of 3 as μ increases. These results can be verified by direct calculation (see Problem 6.2).

Another important property of the distribution is that the sum of two independent Poisson variates itself follows the Poisson distribution. For the m.g.f. of the distribution is

$$M(t) = E(e^{tX}) = e^{-\mu} \sum_{x=0}^{\infty} e^{tx} \frac{\mu^x}{x!} = e^{-\mu} \sum_{x=0}^{\infty} \frac{(\mu e^t)^x}{x!} = e^{-\mu}e^{\mu e^t}.$$

If X and Y are independent Poisson variates with means μ and ν respectively then the m.g.f. of their sum is

$$e^{-\mu}e^{\mu e^t}e^{-\nu}e^{\nu e^t} = e^{-(\mu+\nu)}e^{(\mu+\nu)e^t}$$

which is the m.g.f. of a Poisson variate with mean $\mu+\nu$. For example, we shall see later that the number of radioactive disintegrations in a fixed time is usually a Poisson variate; if we measure the pooled output from two or more sources of radioactive disintegration it will also follow this distribution.

The Poisson distribution was first derived in 1837 by S. D Poisson in a book which purported to show how probability theory could be applied to the decisions of juries, but it attracted little attention until the publication in 1898 of a remarkable little book, *Das Gesetz der kleinen Zahlen*, in which the author, von Bortkiewicz, showed how the distribution could be used to explain statistical regularities in the occurrence of rare events. As examples von Bortkiewicz considered the numbers of suicides of children in different years, the numbers of suicides of women in different states and years, the numbers of accidental deaths in different years in 11 trade associations and the numbers of deaths from horse kicks in the Prussian Army in different years. The last example provided the most extensive data and has become a classical example of the Poisson distribution.

TABLE 13

The annual numbers of deaths from horse kicks in 14 Prussian army corps between 1875 and 1894

Number of deaths	Observed frequency	Expected frequency
0	144	139
1	91	97
2	32	34
3	11	8
4	2	1
5 and over	0	0
Total	280	280

Von Bortkiewicz extracted from official records the numbers of deaths from horse kicks in 14 army corps over the twenty-year period 1875-1894, obtaining 280 observations in all. He argued that the chance that a particular soldier should be killed by a horse in a year was extremely small, but that the number of men in a corps was very large so that the number of deaths in a corps in a year should follow the Poisson distribution. The data are given in Table 13. The total number of deaths is 196, so that the number of deaths per corps per year is $196/280 = \cdot 70$, which can be used as an estimate of μ. The expected numbers in the third column have been calculated by substituting ·70 for μ in the Poisson formula and then multiplying by 280; that is to say, they are the successive terms in the series

$$280e^{-0\cdot7}\left(1,\ 0\cdot7,\ \frac{0\cdot7^2}{2},\ \frac{0\cdot7^3}{6},\ \frac{0\cdot7^4}{24}\ \text{and so on}\right).$$

They are in good agreement with the observed numbers. The variance of the distribution is ·75 which is nearly equal to the mean.

The examples cited by von Bortkiewicz aroused considerable interest by their rather startling nature but they may have given rise to a misleading impression that the Poisson distribution is only a statistical curiosity. The first biological application of the distribution was given by 'Student' (1907) in his paper on the error of counting yeast cells in a haemocytometer, although he was unaware of the work of Poisson and von Bortkiewicz and derived the distribution afresh. (' Student ' was the pen-name used by W. S. Gosset who was employed by Messrs Guinness in Dublin; employees of the brewery were not allowed to publish under their own name. The source of his interest in yeast cells is obvious.) A haemocytometer is essentially an instrument for counting the number of cells in a cell-suspension. A known volume of the suspension is spread uniformly over a slide which is ruled into a large number of squares. The numbers of cells in the individual squares can then be counted under a microscope. It is usual to count only a sample of the squares, and this of course may introduce an error since the number of cells will vary from square to square. 'Student' was concerned to calculate the magnitude of the error introduced in this way. He argued that the number of cells in a square should follow a Poisson distribution, although he did not know it by that name, since the chance that a particular cell will fall in a particular square is small (1 in 400), but there are a large number of cells which may do so. He tested this theory on four distributions which had been counted over the whole 400 squares of the haemocytometer. The results are given in Fig. 17 which gives a good idea of the change of the shape of the distribution as the mean increases. The agreement with the expected numbers, calculated in the same way as before, is excellent. It will also be observed that in all cases the variance is approximately equal to the mean.

Another well-known application of the Poisson distribution is found in the disintegration of radioactive substances. In a

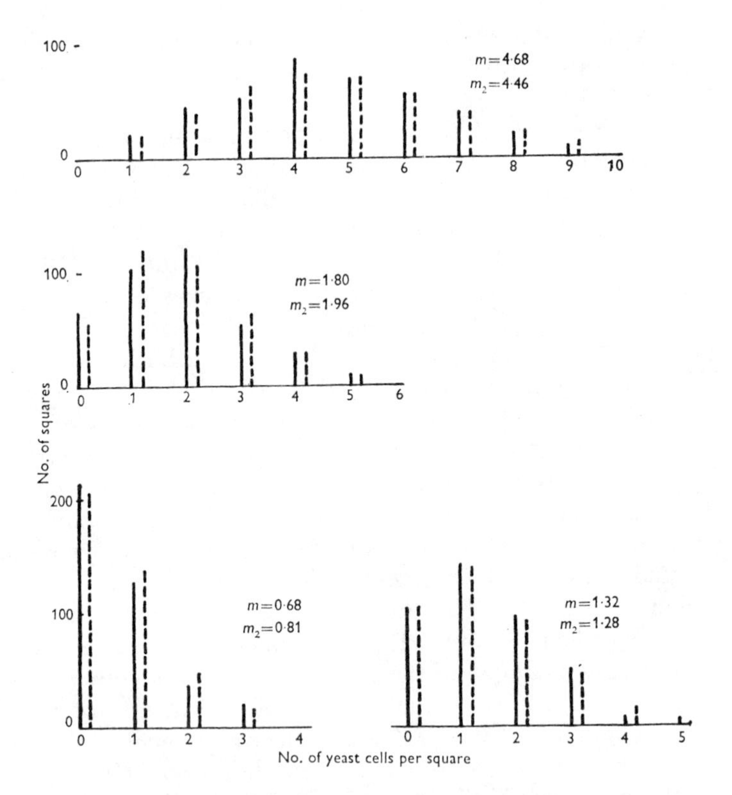

FIG. 17. Data from 'Student' on the distribution of the numbers of yeast cells counted over the 400 squares of the haemocytometer. ——— Observed number of squares. - - - - - Expected number of squares

classical experiment Rutherford, Geiger and Bateman (1910) counted the number of α-particles emitted by a film of polonium in 2608 successive intervals of one-eighth of a minute. They argued that this should follow a Poisson distribution, although again they did not know it by this name and were unaware of any previous work on the subject, since there is a very large number of atoms each of which has a very small chance of disintegrating in one-eighth of a minute. Their data are given in Table 14 and can be seen to be in excellent agreement with theory. The mean of the distribution is 3·87 and the variance 3·69.

TABLE 14

The distribution of the number of α-particles emitted by a film of polonium in 2608 intervals of one-eighth of a minute

	No. of intervals	
No. of α-particles	Observed	Expected
0	57	54
1	203	210
2	383	407
3	525	525
4	532	508
5	408	394
6	273	254
7	139	140
8	45	68
9	27	29
10	10	11
11	4	4
12	0	1
13	1	1
14	1	1
Over 14	0	0
Total	2608	2608

The Poisson distribution has thus a wide variety of applications in situations where events happen randomly in space or in time. The occasions on which the distribution fails to fit the facts are almost as interesting as those on which it holds since deviations from theory indicate some degree of non-randomness whose causes can be investigated. Table 15 shows the distribution of the number of accidents in 5 weeks among a group of women working on high explosive shells in a munitions factory during the 1914-1918 War (Greenwood and Yule, 1920). One would expect this distribution to follow

Poisson's law if an accident can be regarded as a random event happening to a woman, since for each woman there is a very small probability that she will have an accident in a short interval of time, such as five minutes, but there is a large number of such intervals in 5 weeks. (This argument may seem rather artificial; an alternative derivation of the Poisson distribution in cases like this is given in the next section.) It is clear, however, that there are too many women in the extreme groups, with either no accident or with three or more accidents, and too few with only one accident; this is reflected in the fact that the variance is ·69 compared with a mean of ·47. Greenwood and Yule explained the discrepancies by

TABLE 15

The distribution of the number of accidents among a group of women in a munitions factory

No. of accidents	0	1	2	3	4	5	Over 5	Total
Observed no. of women	447	132	42	21	3	2	0	647
Predicted no. of women	406	189	44	7	1	0	0	647

supposing that the group of women was not homogeneous but that they differed in their accident proneness, that is to say in the probability that they will have an accident in a short interval of time; this would obviously increase the spread of the distribution. (See Problem 8.9.)

Another example, which demonstrates the way in which deviations from Poisson's law may be generated, is quoted by Bartlett (1960) from a paper by Bowen (1947). This paper gives the distribution of adult beet leaf-hoppers on 360 beet in a field on two dates (20th May and 26th August) in 1937. On the first date the leaf-hoppers have just arrived and have distributed themselves at random according to Poisson's law; whereas the second count represents the second generation which has developed on the same beet from the individuals in the first generation, and cannot be represented by a Poisson distribution owing to the clumping effect.

Most deviations from Poisson's law have the effect of increasing the numbers in the extreme classes, which is reflected in an increase in the variance. An interesting example of a decrease in the scatter is given by Berkson *et al.* (1935) who investigated the distribution of blood cells in a haemocytometer. They found that, while the distribution of white blood cells followed Poisson's law very well, there were slightly too few red blood cells in the extreme classes; in the latter case the variance was consistently about 85 per cent of the mean. They explained this fact by the observation that while they were settling on the slide the red blood cells could be seen to collide and to repel each other into neighbouring squares; this would have the effect of evening out their distribution and so of reducing the variance slightly. This behaviour was not observed in white blood cells or yeast cells, which follow the Poisson distribution, since they are rather amorphous and not elastic like red blood cells.

The Exponential Distribution

A continuous, positive random variable is said to follow the exponential distribution if its probability density function is

$$f(x) = \lambda e^{-\lambda x}, \quad x \geqslant 0.$$

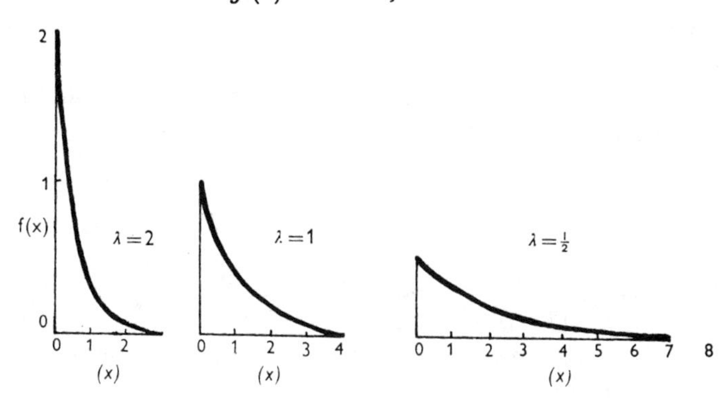

FIG. 18. Density function of the exponential distribution for different values of λ

The cumulative probability function is

$$F(x) = Prob\ [X \leqslant x] = \int_0^x \lambda e^{-\lambda u} du = [-e^{-\lambda u}]_0^x = 1 - e^{-\lambda x}.$$

This function increases from 0 to 1, as it should do, as x increases from 0 to ∞. The distribution is illustrated in Fig. 18 for different values of λ. It will be seen that its shape is always the same, but that its scale *increases* as λ *decreases.*

The mean of the distribution is

$$\int_0^\infty \lambda x e^{-\lambda x} dx.$$

If we make the substitution $y = \lambda x$ this integral becomes

$$\frac{1}{\lambda}\int_0^\infty y e^{-y} dy = \frac{1}{\lambda}\Big[-(y+1)e^{-y}\Big]_0^\infty = \frac{1}{\lambda}.$$

The mean thus increases as λ decreases; such behaviour might be anticipated from Fig. 18. The median is the value of x for which

$$F(x) = 1 - e^{-\lambda x} = \tfrac{1}{2}$$

and is thus $-\log \tfrac{1}{2}/\lambda = \cdot 693/\lambda$. The mode of the distribution is at the origin, and is thus not a sensible measure of the 'centre' of the distribution. It will be noticed that the median is very nearly two-thirds of the way between the mode and the mean.

The higher moments are most easily found from the moment generating function which is

$$M(t) = \lambda \int_0^\infty e^{tx} e^{-\lambda x} dx = \lambda \int_0^\infty e^{-(\lambda - t)x} dx = \frac{\lambda}{\lambda - t}.$$

The variance is $1/\lambda^2$, so that the standard deviation is equal to the mean. The third moment is $2/\lambda^3$, so that the skewness is $+2$. It is clear from the figure that the distribution is highly skew to the right.

The exponential distribution arises naturally as the distribution of waiting times in many situations in which events can be regarded as occurring randomly in time. Let us consider radioactive disintegration as an example. If, on the average, there are λ disintegrations per second, then the average number of disintegrations in t seconds must be λt. Hence, if disintegrations occur at random, the number of disintegrations in t seconds will follow a Poisson distribution with mean λt; in particular, the probability that there will be no disintegrations in t seconds is $e^{-\lambda t}$. But this is the same as the probability that we shall have to wait more than t seconds before the first disintegration occurs. If, therefore, we write T for the waiting time before the occurrence of the first disintegration, then

$$Prob\ [T>t] = e^{-\lambda t}$$

and

$$Prob\ [T\leq t] = 1-e^{-\lambda t}.$$

But the latter is the cumulative probability function of the distribution of waiting times; it follows that T follows an exponential distribution with parameter λ.

TABLE 16

Distribution of the time interval between successive disintegrations of thorium

Time interval (seconds)	Observed frequency	Expected frequency
0-½	101	109
½-1	98	93
1-2	159	149
2-3	114	110
3-4	74	80
4-5	48	59
5-7	75	76
7-10	59	54
10-15	32	28
15-20	4	6
20-30	2	2
Over 30	0	0
Total	766	766

In practice, one has records of the times, t_1, t_2, t_3, t_4 and so on, at which successive disintegrations occurred. If one considers each disintegration in turn as the origin of time it will be seen that the differences between the times of successive disintegrations, t_1, t_2-t_1, t_3-t_2, t_4-t_3 and so on, should follow the exponential distribution. A typical frequency distribution, taken from a paper by Marsden and Barratt (1911) on the radioactive disintegration of thorium, is given in Table 16. There were altogether 766 disintegrations in 2496 seconds, so that λ, the mean number of disintegrations per second, can be estimated by $766/2496 = \cdot 3069$. (It may be noted in passing that the average time between

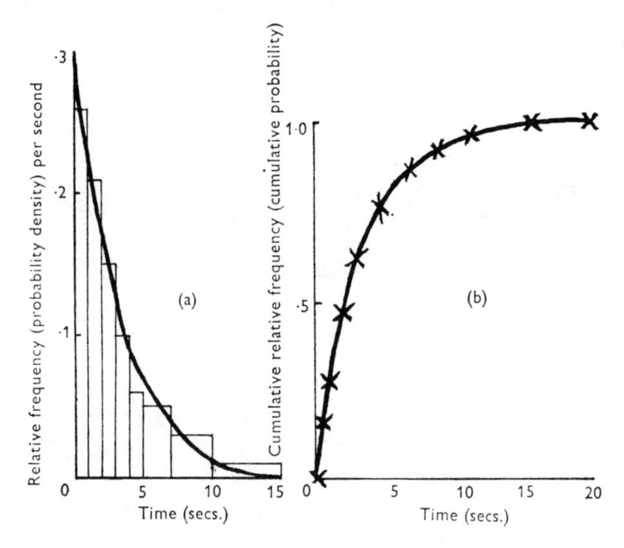

FIG. 19. (a) The histogram and (b) the cumulative relative frequency of the data in Table 16 together with the corresponding theoretical curves

successive disintegrations must be $2496/766 = 1/\cdot 3069 = 3\cdot 26$ seconds.) The expected numbers have been calculated by substituting this value for λ in the cumulative probability function. For example,

$$\begin{aligned} Prob\ [1 < T \leqq 2] &= Prob\ [T \leqq 2] - Prob\ [T \leqq 1] \\ &= F(2) - F(1) = e^{-\lambda} - e^{-2\lambda} = e^{-\cdot 3069} - e^{-\cdot 6138} = \cdot 195 \end{aligned}$$

so that the expected number of intervals between 1 and 2 seconds is $766 \times \cdot 195 = 149$. Agreement with the observed numbers is good. The histogram and the cumulative frequency function are shown in Fig. 19 together with the theoretical curves for $\lambda = \cdot 3069$. Note the importance of using the relative frequency *per second* in the histogram when the class intervals are different.

Fig. 20 shows another example of an exponential distribution from a paper by Fatt and Katz (1952) on the spontaneous miniature potentials arising from the end-plates of resting muscle fibres. The fact that the time interval between successive discharges follows an exponential distribution is strong evidence that they are occurring at random; the authors

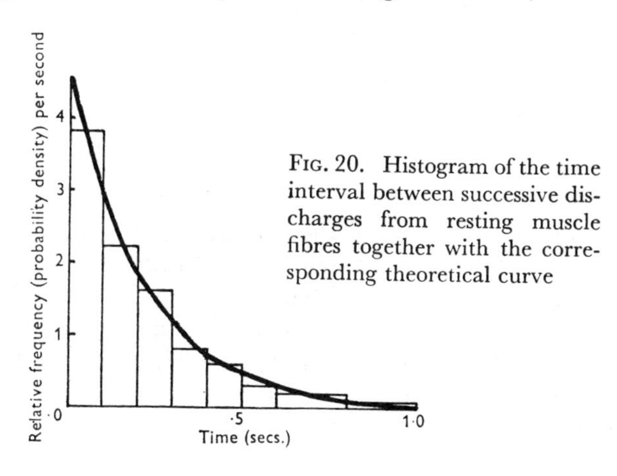

Fig. 20. Histogram of the time interval between successive discharges from resting muscle fibres together with the corresponding theoretical curve

suggest that they may be due to thermal agitation of ions across the nerve membranes.

In deriving the exponential distribution for waiting times we relied on the fact that the number of events in a fixed interval of time follows the Poisson distribution. The distribution can also be derived by an independent argument which may be illuminating. We suppose that in a short interval of time, Δt, there is a chance $\lambda\Delta t$ that an event will occur; if Δt is very small then the chance that more than one event will occur is negligible, so that the chance that no event occurs is $1-\lambda\Delta t$. It will be assumed that the chance of an event occurring in a short time does not depend on how many events have already occurred (this is what we mean by saying that events occur *at random*). If we write $P_0(t)$ for the probability that no event occurs before time t, then the probability that no event occurs before time $(t+\Delta t)$ is the probability that no event occurs before time t multiplied by the probability that no event occurs between time t and time $(t+\Delta t)$; in symbols,

$$P_0(t+\Delta t) = P_0(t)[1-\lambda\Delta t]$$

or

$$\frac{P_0(t+\Delta t)-P_0(t)}{\Delta t} = -\lambda P_0(t).$$

If we let Δt tend to zero, this gives rise to the differential equation

$$\frac{dP_0(t)}{dt} = -\lambda P_0(t)$$

whose solution is

$$P_0(t) = e^{-\lambda t}.$$

(The initial condition is $P_0(0) = 1$.) Now $P_0(t) = 1-F(t)$, from which it follows that the waiting time until the first event occurs follows an exponential distribution.

This argument can be extended to derive the Poisson distribution for the number of events occurring in a fixed time t. Let us write $P_n(t)$ for the probability that exactly n events have occurred by time t. Now n events can occur by time $(t+\Delta t)$ in two ways: (1) n events occur by time t, and no events between times t and $t+\Delta t$, or (2) $(n-1)$ events occur by time t, and one event between times t and $t+\Delta t$ (we are assuming that $n \geqq 1$); hence

$$P_n(t+\Delta t) = P_n(t)[1-\lambda\Delta t]+P_{n-1}(t)\lambda\Delta t$$

or

$$\frac{P_n(t+\Delta t)-P_n(t)}{\Delta t} = \lambda[P_{n-1}(t)-P_n(t)].$$

If we let Δt tend to zero, this gives rise to the differential equation

$$\frac{dP_n(t)}{dt} = \lambda[P_{n-1}(t)-P_n(t)].$$

This equation can now be solved successively for $n = 1, 2, 3$ and so on. Thus we already know that $P_0(t) = e^{-\lambda t}$; hence

$$\frac{dP_1(t)}{dt} = \lambda[P_0(t)-P_1(t)] = \lambda e^{-\lambda t}-\lambda P_1(t)$$

whose solution is

$$P_1(t) = \lambda t e^{-\lambda t}$$

with the initial condition $P_1(0) = 0$. We can now write down a differential equation for $P_2(t)$, and so on. It will be found that

$$P_n(t) = \frac{(\lambda t)^n e^{-\lambda t}}{n!}$$

which is the Poisson distribution with $\mu = \lambda t$.

This argument is typical of the type of argument used in the study of stochastic processes, or random processes occurring in time. The very simple model which we have considered can be modified and extended in several ways. For example, if we are studying the growth of a natural population it might be reasonable to suppose that if n individuals exist at time t, then the probability that a new individual will be born before time $t+\Delta t$ is $\lambda n \Delta t$, where λ is the birth-rate per individual; this stochastic process is called a birth process. We might also wish to take into account the probability that an individual will die, which would give rise to a birth and death process. As additional complications we might suppose that the birth and death rates per individual are density dependent, that is to say are functions of n, or depend on the age of the individual. In general it can be said that the more complicated the model, the more likely it is to fit the biological or physical situation but the less tractable it becomes mathematically; as in other branches of applied mathematics a compromise has to be sought between these two factors. The theory of stochastic processes has found a wide variety of applications such as cascade processes in physics, the formation of queues, the spread of epidemics, the growth of bacterial populations, telephone engineering and so on. The interested reader should consult one of the books devoted to the subject, such as Bailey (1964).

Exercises

6.1. In a table of random numbers, digits are arranged in groups of four. One proposed test of randomness is the *poker* test in which the groups are classified as

(*i*) all four digits the same
(*ii*) three digits the same, and one different
(*iii*) two of one digit and two of another
(*iv*) two of one digit and two different digits
(*v*) all four different

and the observed frequencies compared with expectation. Find the probabilities of each of these events. (P. P. P., Hilary, 1963.]

6.2. When Mendel crossed a tall with a dwarf strain of pea and allowed the hybrids to self-fertilise he found that $\frac{3}{4}$ of the offspring were tall and $\frac{1}{4}$ dwarf. If 9 such offspring were observed, what is the chance (*a*) that less than half of them would be tall?; (*b*) that all of them would be tall?

6.3. Find the mean and variance of the observed frequency distribution and of the theoretical probability distribution in Table 11 on p. 82 directly, using the exact probabilities in fractional form for the latter; compare these values with those obtained from the formula for the binomial distribution with $n=5$ and $P=\frac{1}{2}$.

6.4. Weldon threw 12 dice 4096 times, a throw of 4, 5 or 6 being called a success, and obtained the following results:

No. of successes	0	1	2	3	4	5	6	7	8	9	10	11	12	Total
Frequency	0	7	60	198	430	731	948	847	536	257	71	11	0	4096

(*a*) Calculate the mean of the distribution and compare it with that of a binomial distribution with $P=\frac{1}{2}$; find the observed proportion of successes per throw (p).

(*b*) Calculate the variance of the distribution and compare it with that of a binomial distribution (*i*) with $P=\frac{1}{2}$; (*ii*) with $P=p$.

(*c*) Fit a binomial distribution, (*i*) with $P=\frac{1}{2}$, (*ii*) with $P=p$.

6.5. Evaluate the variance of the sex ratios in Table 1 on p. 3 (*a*) in the Regions of England, (*b*) in the rural districts of Dorset, and compare them with the theoretical variances of a proportion (*a*) with $n=100{,}000$, (*b*) with $n=200$.

6.6. The following is one of the distributions of yeast cells observed by 'Student'. Fit a Poisson distribution to it.

No. of yeast cells	0	1	2	3	4	5	6	Total
No. of squares with this no. of cells in it	103	143	98	42	8	4	2	400

6.7. The frequency of twins in European populations is about 12 in every thousand maternities. What is the probability that there will be no twins in 200 births, (*a*) using the binomial distribution, (*b*) using the Poisson approximation?

6.8. When a bacterial culture is attacked by a bacterial virus almost all the bacteria are killed but a few resistant bacteria survive. This resistance is hereditary, but it might be due *either* to random, spontaneous mutations occurring before the viral attack, *or* to mutations directly induced by this attack. In the latter case the number of resistant bacteria in different but similar cultures should follow the Poisson distribution, but in the former case the variance should be larger than the mean since the number of resistant bacteria will depend markedly on whether a mutation occurs early or late in the life of the culture. In a classical experiment Luria & Delbruck (1943) found the following numbers of resistant bacteria in ten samples from the same culture: 14, 15, 13, 21, 15, 14, 26, 16, 20, 13, while they found the following numbers in ten samples from different cultures grown in similar conditions: 6, 5, 10, 8, 24, 13, 165, 15, 6, 10. Compare the mean with the variance in the two sets of observations. Comment.

6.9. Calculate the mean and the standard deviation of the distribution in Table 16 on p. 99 and show that they are nearly equal.

6.10. If traffic passes a certain point at random at a rate of 2 vehicles a minute, what is the chance that an observer will see no vehicles in a period of 45 seconds?

Problems

6.1. The rth factorial moment, $\mu_{[r]}$, is the Expected value of $X(X-1) \ldots (X-r+1)$. Find the rth factorial moment of the binomial distribution (*a*) by direct calculation, (*b*) from the p.g.f. and hence find its mean, variance, skewness and kurtosis. (Cf. Problem 5.9.)

6.2. Do the same calculations for the Poisson distribution.

6.3. If X and Y are independent Poisson variates with means μ and ν, show that

$$\text{Prob}\,[X+Y=c] = \sum_{i=0}^{c} \text{Prob}\,[X=i \;\&\; Y=c-i] = \frac{e^{-(\mu+\nu)}(\mu+\nu)^c}{c!}.$$

6.4. Suppose that a suspension of cells contains μ cells per c.c. so that the number of cells in 1 c.c. follows a Poisson distribution with mean μ; and suppose that the probability that a cell will survive and multiply to form a visible colony is P, so that the number of colonies formed from a fixed number of cells follows a binomial distribution. Show that the number of colonies formed from 1 c.c. of the suspension follows a Poisson distribution with mean μP.

6.5. If skylarks' nests are distributed at random over a large area of ground show that the square of the distance of a nest from its nearest neighbour follows the exponential distribution. (This fact has been used to test whether there is any territorial spacing of nests.)

6.6. In a pure birth process it is assumed that an individual has a probability $\lambda\Delta t$ of splitting into two during a short time interval Δt, so that, if there are n individuals at time t, the probability that a new individual will be born before time $t+\Delta t$ is $\lambda n\Delta t$. Show that $P_n(t)$, the probability that there are n individuals at time t, satisfies the differential equation

$$\frac{dP_n(t)}{dt} = -n\lambda P_n(t)+(n-1)\lambda P_{n-1}(t) \qquad n \geqq 1$$

and hence show that if there is one individual at time $t=0$,

$$P_n(t) = e^{-\lambda t}(1-e^{-\lambda t})^{n-1}.$$

6.7. Suppose that an urn contains N balls of which R are red and N-R black, and that n balls are drawn from the urn at random. If the balls are replaced after they are drawn the number of red balls in the sample will follow a binomial distribution with $P=R/N$ but if they are not replaced the proportion of red balls in the urn will change from one trial to the next and the distribution will be modified. Show that, under sampling without replacement, the number of red balls in the sample follows the *hypergeometric distribution*

$$P(x) = \frac{\binom{R}{x}\binom{N-R}{n-x}}{\binom{N}{n}}, \quad x=0, 1, 2, \ldots, \text{smaller of } R \text{ or } n$$

where

$$\binom{R}{x} = \frac{R!}{x!(R-x)!}$$

is the number of combinations of x objects out of R.

Evaluate and compare the binomial and hypergeometric distributions, appropriate to sampling with and without replacement, for a sample of 3 balls drawn from an urn containing 4 red and 6 black balls.

6.8. Let $Z_i=0$ or 1 acording as the ith ball drawn in sampling without replacement is black or red. Clearly the probability that the ith ball is red is $P=R/N$ and the probability that both the ith and the jth balls are red is $R(R-1)/N(N-1)$. Find $E(Z_i)$, $V(Z_i)$ and $\text{Cov}(Z_i, Z_j)$ and hence show that

$$E(X)=nP$$

$$V(X)=nPQ\left(1-\frac{n-1}{N-1}\right).$$

Compare these results with those for the binomial distribution. Verify them directly for the distribution evaluated in the previous problem.

6.9. In a series of independent trials with a constant probability of success show that the probability that there will be exactly x failures before the first success is

$$P(x) = Q^x P, \quad x = 0, 1, 2, \ldots$$

This distribution is called the geometric distribution. It has applications as the distribution of waiting times for events occurring at discrete intervals of time and is thus the discrete analogue of the exponential distribution.

Find the probability generating function of the distribution. Hence find an expression for $\mu_{[r]}$ and use this expression to find its mean, variance and skewness.

6.10. Suppose that members of an effectively infinite population can be classified not just into two but into k classes and that the probability that a member belongs to the ith class is P_i. Show that if n members are chosen at random the probability that exactly x_1 will belong to the first class, x_2 to the second class and so on is

$$P(x_1, x_2, \ldots, x_k) = \frac{n!}{x_1!x_2!\ldots x_k!} P_1^{x_1} P_2^{x_2} \ldots P_k^{x_k}$$

This distribution, which is an obvious extension of the binomial distribution, is called the *multinomial distribution*. The probabilities are the terms of the expansion of $(P_1+P_2+\ldots+P_k)^n$.

Find the Covariance of X_i and X_j (*a*) directly, (*b*) from the formula $V(X_i+X_j) = V(X_i)+V(X_j)+2 \text{ Cov } (X_i, X_j)$.

Chapter 7

THE NORMAL DISTRIBUTION

Order in Apparent Chaos—*I know of scarcely anything so apt to impress the imagination as the wonderful form of cosmic order expressed by the " Law of Frequency of Error ". The law would have been personified by the Greeks and deified, if they had known of it. It reigns with serenity and in complete self-effacement amidst the wildest confusion. The huger the mob, and the greater the apparent anarchy, the more perfect is its sway. It is the supreme law of Unreason.*

Francis Galton: *Natural Inheritance* (1889)

On a donc fait une hypothèse, et cette hypothèse a été appelée loi des erreurs. Elle ne s'obtient pas par des déductions rigoureuses. . . . " Tout le monde y croit cependant," me disait un jour M. Lippmann, " car les expérimentateurs s'imaginent que c'est un théorème de mathématiques, et les mathématiciens que c'est un fait expérimental ".

H. Poincaré : *Calcul des probabilités* (1896)

The normal distribution was discovered in 1733 by the Huguenot refugee Abraham de Moivre as an approximation to the binomial distribution when the number of trials is large. The distribution is more usually associated with the name of Gauss who derived it in 1809, in his *Theoria motus corporum coelestium*, as the law of errors of observations, with particular reference to astronomical observations. Gauss argued that if a number of observations of the same quality are taken, each subject to random error, the natural way of combining them is to find their mean. He then showed that this is the best way of combining them (by which he meant, roughly speaking, the method of maximum likelihood) only if the errors follow what is now called the normal distribution, from which he concluded that they must do so. This rather curious argument does not carry much conviction today, but Gauss was also able to show empirically that errors in astronomical observations do follow the normal curve with considerable accuracy.

The real reason for the importance of the normal distribution lies in the central limit theorem which states that the sum of a large number of independent random variables will be approximately normally distributed almost regardless of their individual distributions; any random variable which can be regarded as the sum of a large number of small, independent contributions is thus likely to follow the normal distribution approximately. This theorem was first proved by Laplace in 1812 but his proof was so complicated as to be almost incomprehensible and the significance of the theorem was not realised until Quetelet demonstrated in 1835 that the normal curve described the distribution not only of errors of measurement but also of biological variables such as human height. It was soon discovered that many other biometrical variables followed this distribution; it was at this time that it acquired the name of the *normal* distribution, which any well-behaved variable ought to follow.

There is no doubt that the importance of the normal distribution was exaggerated in the first flush of enthusiasm, typified in Galton's encomium quoted at the beginning of the chapter. Later statisticians took a more sober approach, illustrated in the quotation by Poincaré, and recognised that many distributions are not normal. Nevertheless the central limit theorem has ensured for this distribution its central place in statistical theory. In this chapter we shall first describe the mathematical properties of the normal distribution, and then discuss the central limit theorem and its applications.

Properties of the Normal Distribution

A continuous random variable is said to be normally distributed with mean μ and variance σ^2 if its probability density function is

$$f(x) = \frac{1}{\sigma\sqrt{2\pi}} e^{-\frac{1}{2}(x-\mu)^2/\sigma^2}, \quad -\infty < x < \infty$$

This formula defines a family of distributions depending on the two parameters μ and σ. A change in the mean, μ, shifts the distribution bodily along the x-axis; a change in the standard deviation, σ, flattens it or compresses it while leaving its centre in the same position. The distribution is illustrated in Fig. 21. Its shape has been likened to that of a bell and a cocked hat.

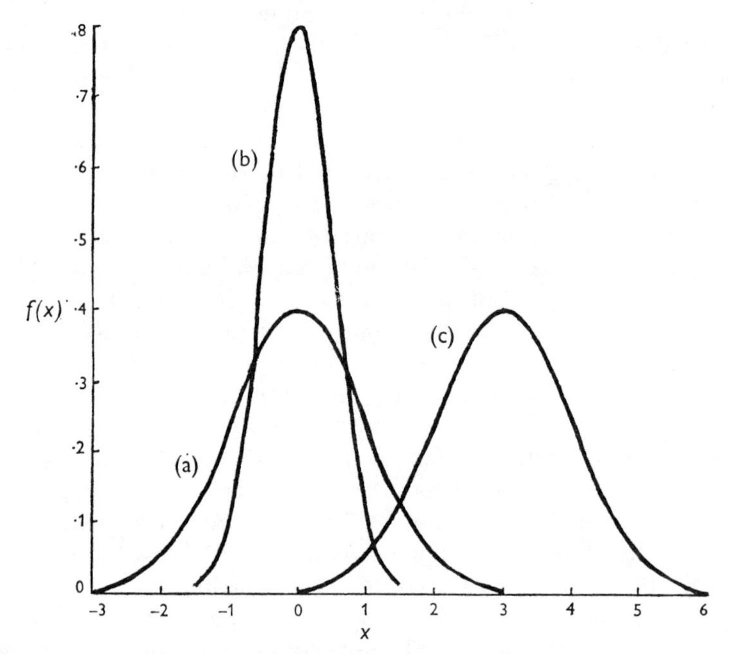

FIG. 21. The density function of the normal distribution with (a) $\mu = 0$, $\sigma = 1$; (b) $\mu = 0$, $\sigma = \frac{1}{2}$; (c) $\mu = 3$, $\sigma = 1$

It is proved in the Appendix to this chapter that the area under the density function is unity; the factor $1/\sigma\sqrt{2\pi}$ ensures that this is so. The really important part of the formula is the exponential of $-\frac{1}{2}\left(\frac{x-\mu}{\sigma}\right)^2$. This implies that the probability density depends only on the distance of x from its mean value, μ, expressed as a proportion of the standard deviation, σ. The density function attains its maximum value of $1/\sigma\sqrt{2\pi} = \cdot 4/\sigma$ when $x = \mu$ and falls off symmetrically on each side of μ. It follows immediately that the mean, median and mode of the distribution are all equal to μ and that the odd moments about the mean, and in particular the third moment, μ_3, are all zero.

The even moments are most easily found by computing the moment generating function:

$$M(t) = \frac{1}{\sigma\sqrt{2\pi}} \int_{-\infty}^{\infty} e^{tx - \frac{1}{2}(x-\mu)^2/\sigma^2} dx.$$

If we expand the exponent in this integral and then complete the square it will be found that it can be expressed in the form:

$$M(t) = \frac{e^{\mu t + \frac{1}{2}\sigma^2 t^2}}{\sigma\sqrt{2\pi}} \int_{-\infty}^{\infty} e^{-\frac{1}{2}(x-\mu-\sigma^2 t)^2/\sigma^2} dx = e^{\mu t + \frac{1}{2}\sigma^2 t^2}.$$

To find the moments about the mean we must consider

$$e^{-\mu t} M(t) = e^{\frac{1}{2}\sigma^2 t^2} = 1 + \tfrac{1}{2}\sigma^2 t^2 + \tfrac{1}{4}\frac{\sigma^4 t^4}{2!} + \tfrac{1}{8}\frac{\sigma^6 t^6}{3!} + \ldots$$

Now μ_r is the coefficient of t^r multiplied by $r!$. It follows that $\mu_2 = \sigma^2$, as was stated above, and that $\mu_4 = 3\sigma^4$. The kurtosis is therefore 3, which is the reason why this is taken as the standard kurtosis. The rth moment, if r is even, is

$$\mu_r = 1 \cdot 3 \cdot 5 \ldots (r-1)\sigma^r.$$

If X is normally distributed with mean μ and variance σ^2, then any linear function of X, say $Y = a + bX$, is also normally distributed with the appropriate mean and variance. (The mean is $a + b\mu$ and the variance $b^2\sigma^2$.) This follows from the fact that the formula for the normal density function contains the arbitrary location and scale parameters μ and σ; a change in the scale and origin of the observations will change these parameters but will not alter the shape of the distribution. To prove this result mathematically we first recall that if the m.g.f. of X is $M(t)$ then the m.g.f. of $Y = a + bX$ is $e^{at}M(bt)$. If X is normally distributed with mean μ and variance σ^2 then its m.g.f. is

$$M(t) = e^{\mu t + \frac{1}{2}\sigma^2 t^2}$$

and the m.g.f. of Y is

$$e^{at}M(bt) = e^{(a+b\mu)t+\frac{1}{2}b^2\sigma^2t^2}$$

which is the m.g.f. of a normal variate with the appropriate mean and variance.

In particular, if X is normally distributed with mean μ and variance σ^2, then $Z = \frac{X-\mu}{\sigma}$ follows a normal distribution with zero mean and unit variance. Z is called a *standard* normal variate; it is conventional to use the symbols $\phi(z)$ and $\Phi(z)$ for the density function and cumulative probability function respectively of the standard normal distribution. These functions are tabulated at the end of the book; the distribution of any normal variate can be found from them. For, if X is normally distributed with mean μ and variance σ^2 and if we write $F(x)$ and $f(x)$ for its cumulative probability function and density function, then

$$F(x) = Prob\ [X \leqq x] = Prob\left[\frac{X-\mu}{\sigma} \leqq \frac{x-\mu}{\sigma}\right] = \Phi\left(\frac{x-\mu}{\sigma}\right)$$

and $$f(x) = \frac{dF(x)}{dx} = \frac{1}{\sigma}\,\phi\left(\frac{x-\mu}{\sigma}\right).$$

For example, the intelligence quotient (I.Q.), measured by the Stanford-Binet test, is approximately normally distributed in the general population with a mean of 100 and a standard deviation of about 16. The probability that an individual chosen at random will have an I.Q. less than 120 is therefore

$$Prob\ [X \leqq 120] = Prob\left[\frac{X-100}{16} \leqq 1{\cdot}25\right] = \Phi(1{\cdot}25) = 0{\cdot}8944.$$

The density function at this point is $\phi(1{\cdot}25)/16 = {\cdot}1826/16 = {\cdot}0114$; the proportion of people with I.Q.'s between 118 and 122 is therefore approximately $4 \times {\cdot}0114 = {\cdot}0456$.

Another important property of the normal distribution which follows immediately from consideration of its moment generating function is that the sum of two independent normal variates is itself normally distributed. For suppose that

X and Y are independently and normally distributed with means μ_1 and μ_2 and variances σ_1^2 and σ_2^2 respectively, then the m.g.f. of their sum, $X+Y$, is the product of their m.g.f.'s which is easily found to be the m.g.f. of a normal variate with mean $\mu_1+\mu_2$ and variance $\sigma_1^2+\sigma_2^2$. It follows that any linear function of any number of independent normal variates is normally distributed.

TABLE 17

Frequency distribution of height among 3000 criminals (Macdonell, 1901)

Height (inches)	Observed frequency	Expected frequency (normal distribution)
55	0	0
56	1	1
57	1	2
58	6	7
59	23	20
60	48	48
61	90	103
62	175	187
63	317	293
64	393	395
65	462	458
66	458	455
67	413	390
68	264	287
69	177	182
70	97	99
71	46	46
72	17	19
73	7	6
74	4	2
75	0	1
76	0	0
77	1	0
Total	3000	3001

As an example of a random variable which follows the normal distribution with considerable accuracy Table 17 shows the frequency distribution of the heights of 3000 criminals obtained from the records of Scotland Yard (Macdonell, 1901).

(Height was measured to the nearest eighth of an inch; the class 60, for example, includes all heights from $59\frac{9}{16}$ to $60\frac{9}{16}$ inches.) The constants of the distribution are

mean	65·535 inches
standard deviation	2·557 inches
skewness	·064
kurtosis	3·170.

The skewness and kurtosis are in good agreement with the values of the normal distribution. The Expected frequencies have been obtained by calculating the probabilities of obtaining a normal variate with the same mean and variance in each of the classes, and then multiplying these probabilities by 3000. There is clearly good agreement between the Observed and Expected frequencies. The histogram of the distribution is shown in Fig. 22 together with the normal density function

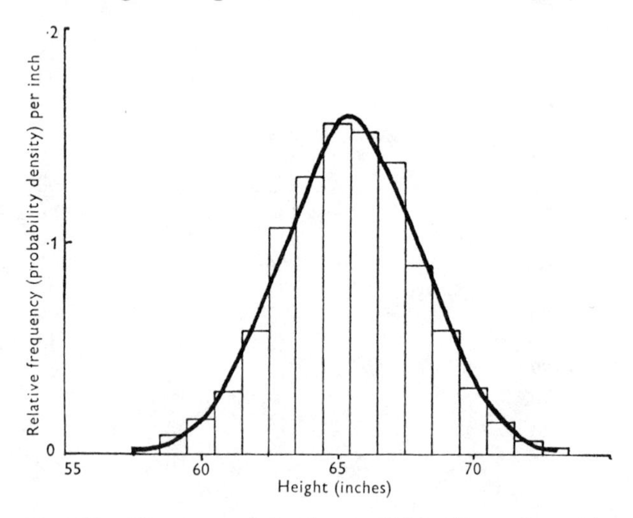

Fig. 22. Histogram of the data in Table 17 together with the corresponding normal curve

with the same mean and variance. The normal distribution of height can be explained by supposing that a man's height is determined by a large number of small, independent, additive factors, both genetic and environmental. We must now consider the central limit theorem which shows why such a variable should follow the normal distribution.

The Central Limit Theorem

The great importance of the normal distribution rests on the central limit theorem which states that the sum of a large number of independent random variables will be approximately normally distributed almost regardless of their individual distributions. Let us therefore consider the distribution of the sum of n independent random variables, $Y = X_1+X_2+\ldots+X_n$. We shall write μ_i, σ_i^2, μ_{3i}, and so on for the mean, variance and higher moments of X_i and μ and σ^2 for the mean and variance of Y, which are of course the sums of the corresponding values of the X_i's. If we write $M_i(t)$ for the moment generating function of $X_i-\mu_i$, then the m.g.f. of $Y-\mu = \Sigma(X_i-\mu_i)$ is the product of the $M_i(t)$'s and so the m.g.f. of the standardised variate $(Y-\mu)/\sigma$ is

$$M^*(t) = \prod_{i=1}^{n} M_i(t/\sigma).$$

Now

$$M_i(t/\sigma) = 1+\frac{\sigma_i^2}{2}\left(\frac{t}{\sigma}\right)^2+\frac{\mu_{3i}}{3!}\left(\frac{t}{\sigma}\right)^3+\ldots$$

When n is large σ^2 will also be large, since it is the sum of the variances of the X_i's; hence σ^3, σ^4 and so on will be large compared with σ^2 and terms containing their reciprocals can be ignored. Furthermore, it is shown in books on calculus that log $(1+h)$ † is very nearly equal to h when the latter is small. Hence, if we take the logarithm of $M^*(t)$, we find that

$$\log M^*(t) = \sum_{i=1}^{n} \log M_i(t/\sigma) \doteqdot \sum_{i=1}^{n} \log\left(1+\tfrac{1}{2}\frac{\sigma_i^2t^2}{\sigma^2}\right)$$

$$\doteqdot \sum_{i=1}^{n} \tfrac{1}{2}\frac{\sigma_i^2t^2}{\sigma^2} = \tfrac{1}{2}t^2.$$

Thus $M^*(t)$ is nearly equal to $e^{\frac{1}{2}t^2}$, which is the m.g.f. of

† Unless otherwise stated, logarithms are understood to be natural logarithms to the base e.

a standard normal variate. It follows that Y is nearly normally distributed with mean μ and variance σ^2.

The assumption that all the random variables possess moment generating functions may not be true. A similar proof can be constructed with characteristic functions which always exist. It can be shown in this way that the sum of a large number of independent random variables will tend to the normal form provided that the variances of all the distributions are finite and provided that, essentially, these variances are of the same order of magnitude. A distribution must be of a very extreme type for its variance to be infinite, although such distributions exist; the Cauchy distribution, which has been considered briefly in Problem 3.4 and which will be met later in the guise of a t distribution with one degree of freedom, is the standard example of a distribution with an infinite variance. It can be shown that the sum of any number of Cauchy variates itself follows the Cauchy distribution, so that this forms an exception to the central limit theorem. Such extreme distributions are, however, very rare and it can be assumed that the central limit theorem will hold under all normal conditions.

The central limit theorem has a wide variety of applications. Firstly, it explains why many distributions in nature are approximately normal; for many random variables can be regarded as the sum of a large number of small independent contributions. For example, it is not unreasonable to suppose that human height is determined by a large number of factors, both genetic and environmental, which are additive in their effects. It must not, however, be thought that all, or even the majority, of naturally occurring distributions are normal, although in many cases this provides as good an approximation as any.

It should be noted that the central limit theorem does not apply unless the factors are independent and additive; the requirement of additivity is particularly important. If the factors are multiplicative, for example, so that the effect produced at any stage is proportional to the size already attained, then the logarithms of the effects will be additive and we should expect the logarithm of the end product to be normally distributed. A variable whose logarithm is normally distributed is said to follow the log-normal distribution; the distribution is skew to the right. For example Sinnott (1937) has shown that the weight of vegetable marrows is log-normally distributed and he interprets this to mean that the genes controlling weight act in a multiplicative rather than an additive manner. (See Fig. 23.)

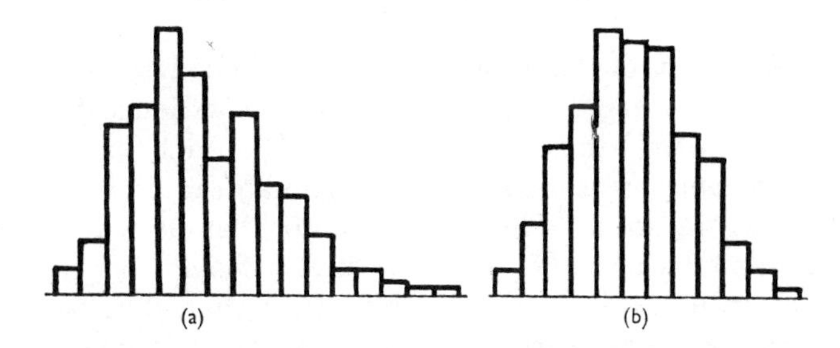

FIG. 23. (a) Segregating population of 244 vegetable marrows plotted in arithmetically equal classes; skewness $+\cdot 487$ (b) The same population plotted in equal logarithmic classes; skewness $-\cdot 057$

Another important application of the central limit theorem is to the binomial distribution. We have already mentioned that the normal distribution was discovered in 1733 by the Huguenot refugee Abraham de Moivre as an approximation to the binomial distribution. De Moivre's proof was rather lengthy and will not be reproduced here. We need merely note that it is a special case of the central limit theorem since the number of successes in n trials can be regarded as the sum of n independent random variables, X_i, which take the values 1 or 0 according as the ith trial results in a success or a failure.

In approximating the binomial probability function by the normal density function with the same mean and variance we are of course approximating a discrete by a continuous function. What we are in fact asserting is that the probability that there will be x successes in n trials is approximately equal to the corresponding normal density function at x which in its turn is nearly equal to the normal integral between $x-\frac{1}{2}$

and $x+\frac{1}{2}$. For example we find from Table 12 on p. 89 that the binomial probability of observing 3 boys in a family of size 8 is $56P^3Q^5 = \cdot 2056$, if we assume that $P = \cdot 5147$. The value of the normal density function with mean $nP = 4\cdot 1176$ and standard deviation $\sqrt{nPQ} = 1\cdot 4136$ is

$$\frac{1}{1\cdot 4136}\,\phi\left(\frac{3-4\cdot 1176}{1\cdot 4136}\right) = \cdot 2065.$$

The accuracy of the approximation depends largely on the size of the variance, nPQ; for when the variance is large only a small probability is concentrated at each mass point and it becomes reasonable to approximate a discrete by a continuous function. The approximation has been found to be quite good when nPQ is as low as 2, although one should be cautious unless it is considerably larger than this, particularly when P is not nearly equal to Q and the binomial distribution is in consequence skew. The approximation holds better in the central portion than in the ends of the distribution. As an example Table 18 shows that the approximation is quite good when $n = 10$, $P = \cdot 2$ and nPQ is only $1\cdot 6$.

TABLE 18

Normal approximation to the binomial distribution for $n = 10$, $P = \cdot 2$

Number of successes	0	1	2	3	4	5	6
Binomial probability	·107	·268	·302	·201	·088	·026	·006
Normal approximation	·090	·231	·315	·231	·090	·019	·002

The chief use of this approximation is in evaluating the 'tails' of the binomial distribution. For example, if one wanted to know how often to expect 27 or more boys in 45 births assuming that the probability of a male birth is ·514, one could work out the answer exactly by summing all the binomial terms from 27 to 45; this would obviously be very tedious. But if we approximate these terms by the normal density function with mean 23·13 and standard deviation 3·353, then the required probability is approximately the integral of this normal distribution from $26\frac{1}{2}$ to infinity which is

$$1-\Phi\left(\frac{26\cdot5-23\cdot13}{3\cdot353}\right) = \cdot1574.$$

The correct probability is ·1579. The integral has been taken from $26\frac{1}{2}$ rather than from 27 upwards to allow for the fact that the binomial distribution is discrete. This correction for continuity is often negligible and may be ignored.

The Poisson distribution can also be approximated by the normal distribution when μ, the counterpart of nPQ, is not too small. Thus if we want to know the probability of observing *less* than 360 radioactive disintegrations in an hour when the true rate of disintegration is 400 per hour, we can argue that the number of disintegrations in an hour is, on our hypothesis, a Poisson variate with mean 400 and is thus approximately normally distributed with mean 400 and standard deviation 20. The required probability is thus nearly

$$\Phi\left(\frac{359\cdot5-400}{20}\right) = \cdot021.$$

We have seen so far how the central limit theorem can be used to explain the frequent occurrence of normal, or nearly normal, distributions in nature, and to justify the use of the normal distributions as an approximation to the binomial and Poisson distributions. The third application of this theorem lies in the theory of sampling distributions which are of great importance in statistical inference and which usually become approximately normal in large samples even though the underlying distribution may be far from normal. We must therefore now explain the important concept of a sampling distribution with particular reference to the sampling distribution of the mean.

Suppose that we have made n observations, $x_1, x_2, \ldots, x_n$, on some random variable and have computed their mean $\bar{x}$; if we make another set of n observations, $x'_1, x'_2, \ldots, x'_n$, and compute *their* mean, $\bar{x}'$, we should not, in general, expect the two means to be identical. If we repeat this procedure a large number of times we will obtain a large number of means, $\bar{x}$, $\bar{x}'$, $\bar{x}''$, and so on which, if we go on long enough, will generate a frequency distribution and eventually a probability distribution. This probability distribution is called the sampling distribution of the mean.

The observed sample of n observations can thus be regarded as one sample drawn at random from the totality of all possible samples. Each of the observations can be thought of in the same way as being drawn at random from the underlying probability distribution and can thus be regarded, if we imagine the sampling procedure to be indefinitely repeated, as independent random variables with this distribution. If the mean and variance of this underlying distribution are μ and σ^2 respectively and if $\bar{x}$ is the mean of the observations, we can therefore write

$$E(n\bar{x}) = E(x_1+x_2+\ldots+x_n) = E(x_1)+E(x_2)+\ldots+E(x_n) = n\mu$$
$$V(n\bar{x}) = V(x_1+x_2+\ldots+x_n) = V(x_1)+V(x_2)+\ldots+V(x_n) = n\sigma^2$$

so that

$$E(\bar{x}) = E(n\bar{x}/n) = E(n\bar{x})/n = \mu$$
$$V(\bar{x}) = V(n\bar{x}/n) = V(n\bar{x})/n^2 = \sigma^2/n.$$

That is to say, in repeated samples of n observations, the sampling distribution of $\bar{x}$ will have mean μ and variance σ^2/n; it is in accordance with common sense that the sampling distribution of $\bar{x}$ should become more closely clustered about μ as n increases. Furthermore, it follows from the central limit theorem that this sampling distribution tends to the normal form as n increases almost regardless of the form of the underlying distribution. If this distribution is itself normal, then so of course is that of $\bar{x}$. In making inferences based on the sample mean it can therefore usually be assumed that $\bar{x}$ is a single observation drawn at random from a normal distribution with mean μ and variance σ^2/n; the assumption of normality will break down only if n is very small or if the underlying distribution is highly abnormal. (It should be observed that the convention of using capital letters for random variables has been abandoned in this paragraph in order to avoid lengthy periphrasis; this practice is usual in the discussion of sampling distributions and should cause no confusion.)

Appendix

The area under the normal curve

We wish to prove that

$$\int_{-\infty}^{\infty} e^{-\frac{1}{2}(x-\mu)^2/\sigma^2} dx = \sigma\sqrt{2\pi}.$$

If we make the substitution $y = (x-\mu)/\sigma$ this is equivalent to showing that

$$A \equiv \int_{-\infty}^{\infty} e^{\frac{1}{2}y^2} dy = \sqrt{2\pi}$$

The square of this integral is

$$A^2 = \int_{-\infty}^{\infty} e^{-\frac{1}{2}x^2} dx \int_{-\infty}^{\infty} e^{-\frac{1}{2}y^2} dy = \int_{-\infty}^{\infty}\int_{-\infty}^{\infty} e^{-\frac{1}{2}(x^2+y^2)} dx dy.$$

This double integral is the volume under the bell-shaped surface $e^{-\frac{1}{2}(x^2+y^2)}$. Now x^2+y^2 is the square of the distance of the point (x, y) from the origin, which we may denote by r^2. Furthermore, the area of a thin annulus or ring at a distance r from the origin and of width dr is $2\pi r dr$ and so the volume of the cylindrical shell with this annulus as base and with height $e^{-\frac{1}{2}r^2}$, so that it just touches the bell-shaped surface, is $e^{-\frac{1}{2}r^2} 2\pi r dr$. Hence the total volume under the surface is

$$A^2 = 2\pi \int_{0}^{\infty} re^{-\frac{1}{2}r^2} dr = 2\pi \Big[-e^{-\frac{1}{2}r^2} \Big]_{0}^{\infty} = 2\pi.$$

This result can also be proved by transforming the double integral from rectangular into polar coordinates if the reader is familiar with this method.

Exercises

7.1. The intelligence quotient (I.Q.) is approximately normally distributed with a mean of 100 and a standard deviation of 16. What is the probability that an individual will have an I.Q. (*a*) less than 90, (*b*) greater than 130, (*c*) between 95 and 105?

7.2. (*a*) What is the probability that the mean I.Q. of a randomly chosen group of 12 people will lie between 95 and 105? (*b*) How large a sample is required to have a chance of 95 per cent that the mean I.Q. of the group will lie between these limits?

7.3. Fit a normal curve to the histogram of the distribution of head breadths calculated in Exercise 3.3 by using a table of the normal density function. (See Exercises 4.6 and 4.12 for the mean and standard deviation of the distribution; use Sheppard's correction for the variance since the grouping is rather coarse.)

7.4. Find the Expected frequencies in the distribution of head breadths in Table 10 on p. 40 on the assumption of normality by using a table of the normal integral.

7.5. From the data in Table 17 on p. 113 estimate the proportion of men with heights (*a*) more than 1 standard deviation, (*b*) more than 2 standard deviations above the mean, and compare the results with the theoretical probabilities on the assumption of normality.

7.6. The reaction times of two motorists *A* and *B* are such that their braking distances from 30 m.p.h. are independently normally distributed. *A* has mean 30 yds and variance 36 yds^2 and *B* has mean 40 yds and variance 64 yds^2. If they both approach each other at 30 m.p.h. on a single track road and first see each other when they are 90 yds apart, what is the probability that they avoid a collision? [P. P. P., Hilary, 1963]

7.7. Given the data of Exercise 2.6 how many fume cupboards would be required to satisfy a group of 50 chemists at least 95 per cent of the time? [Certificate, 1959]

7.8. If the probability of a male birth is ·514 what is the probability that there will be fewer boys than girls in 1000 births? How large a sample must be taken to make the probability of this occurrence less than 5 per cent?

7.9. A chemist estimates that the average demand for tubes of a particular type of toothpaste is 150 per week. Assuming that the weekly demand follows a Poisson distribution, determine the least number of tubes he should have in stock at the beginning of the week if the probability of the stock being exhausted before the end of the week is not to exceed 0·05. [Certificate, 1958]

7.10. In estimating the concentration of cells in a suspension about how many cells must be counted to have a 95 per cent chance of obtaining an estimate within 10 per cent of the true value?

Problems

7.1. Verify that the product of the m.g.f.'s of two normal variates is the m.g.f. of a normal variate with the appropriate mean and variance. Prove by a direct argument not involving the use of the m.g.f. that the sum of two independent normal variates is normally distributed (cf. Problem 3.5).

7.2. Show that the mean deviation of a normal variate is $\sigma\sqrt{\frac{2}{\pi}} = \cdot 798\sigma$.

7.3. If X_1, X_2, ..., X_n are independent random variables with the same distribution, show that the skewness and kurtosis of their sum tend to 0 and 3 respectively as n becomes large (cf. Problem 5.1).

7.4. If X follows the binomial distribution prove directly that the m.g.f. of $(X-\mu)/\sigma$ tends to $e^{\frac{1}{2}t^2}$ as n becomes large. Prove similarly that the Poisson distribution tends to normality as μ becomes large.

7.5. If X and Y are independent Cauchy variates with scale parameter b (see Problem 3.4), show that $X+Y$ is a Cauchy variate with scale parameter $2b$ (cf. Problem 3.5; use partial fractions to integrate.) It follows that this distribution is an exception to the central limit theorem.

7.6. The variates X and Y are normally distributed and are independent of each other. X has mean 10 and unit variance, while Y has mean 2 and variance 4. By using the fact that $Y/X < r$ is almost equivalent to $Y-rX < 0$, show how tables of the standard normal integral could be used to find a close approximation to the probability distribution of Y/X. Indicate how an upper bound could be found for the error in this approximation. [Certificate, 1960]

CHAPTER 8

THE χ^2, t AND F DISTRIBUTIONS

In this chapter we shall consider three distributions which are closely related to the normal distribution. Their importance lies largely in their use as sampling distributions and will become apparent when methods of statistical inference are discussed in succeeding chapters; it is, however, convenient to describe the mathematical properties of these distributions now. Detailed study of the mathematical proofs contained in this chapter may be postponed until a second or subsequent reading.

THE χ^2 DISTRIBUTION

The χ^2 distribution with f degrees of freedom is the distribution of the sum of the squares of f independent standard normal variates, that is to say of

$$Y = Z_1^2 + Z_2^2 + \ldots + Z_f^2$$

where each of the Z's is independently and normally distributed with zero mean and unit variance. This definition generates a family of distributions depending on the number of degrees of freedom, f. It is often convenient to refer to a random variable which follows the χ^2 distribution with f degrees of freedom as a $\chi^2_{[f]}$ variate.

Consider first the mean, variance and skewness of the distribution. The moments about the origin of a $\chi^2_{[1]}$ variate, that is to say of $Y = Z^2$, are

$$\begin{aligned} \mu &= E(Y) = E(Z^2) = 1 \\ \mu_2' &= E(Y^2) = E(Z^4) = 3 \\ \mu_3' &= E(Y^3) = E(Z^6) = 15. \end{aligned}$$

It follows from the rules for converting moments about the origin to moments about the mean that $\mu_2 = 2$ and that

$\mu_3 = 8$. Now a χ^2 variate with f degrees of freedom is the sum of f independent $\chi^2_{[1]}$ variates. It follows from the additive property of the mean and the variance that the χ^2 distribution with f degrees of freedom has mean f and variance $2f$. It is quite easy to show that this additive property holds also for the third moment, so that $\mu_3 = 8f$; the skewness of the distribution is thus $\sqrt{8/f}$. Thus both the mean and the variance of the distribution increase as f increases. When f is small the distribution is highly skew to the right, but the skewness decreases as f increases; when f becomes large the distribution tends to the symmetrical normal form by the central limit theorem.

We will now find the moment generating function, and hence the probability density function, of the χ^2 distribution. The m.g.f. of a $\chi^2_{[1]}$ variate is

$$M(t) = E(e^{tY}) = E(e^{tZ^2}) = \frac{1}{\sqrt{2\pi}} \int_{-\infty}^{\infty} e^{-\frac{1}{2}z^2(1-2t)}dz$$

which is equal to $(1-2t)^{-\frac{1}{2}}$ since apart from this factor it is the area under the normal curve with variance $(1-2t)^{-1}$. The m.g.f. of the χ^2 distribution with f degrees of freedom, that is to say of the sum of f such independent variates, is therefore $(1-2t)^{-\frac{1}{2}f}$. It is easy to verify, by making the substitution $w = (1-2t)y$, that this is the m.g.f. of the density function

$$f(y) = \frac{y^{\frac{1}{2}f-1}e^{-\frac{1}{2}y}}{A(f)} \qquad 0 \leq y < \infty$$

where $A(f)$ is the integral of the numerator from 0 to ∞. (It is clear from its definition that a χ^2 variate is essentially positive.) It is not difficult to show (see Problem 8.2) that

$$A(f) = 1\times3\times5\ldots(f-2)\times\sqrt{2\pi} \qquad \text{when } f \text{ is odd}$$

$$A(f) = 2\times4\times6\ldots(f-2)\times2 = 2^{\frac{1}{2}f}(\tfrac{1}{2}f-1)! \text{ when } f \text{ is even.}$$

Fig. 24 shows the density functions of the χ^2 distribution with different numbers of degrees of freedom. When $f = 1$ the density function is proportional to $e^{-\frac{1}{2}y}/\sqrt{y}$ and therefore increases without bound as y tends to zero; the distribution is thus of an extreme J-shaped type. When $f=2$ the distribution is the exponential distribution with $\lambda = \frac{1}{2}$. When f is greater than 2 the density function rises from zero at the origin to a maximum at $f-2$ and then decreases. The median is very nearly two-thirds of the way between the mode and the mean, that is to say at $f-\frac{2}{3}$. A table of percentage points will be found at the end of the book.

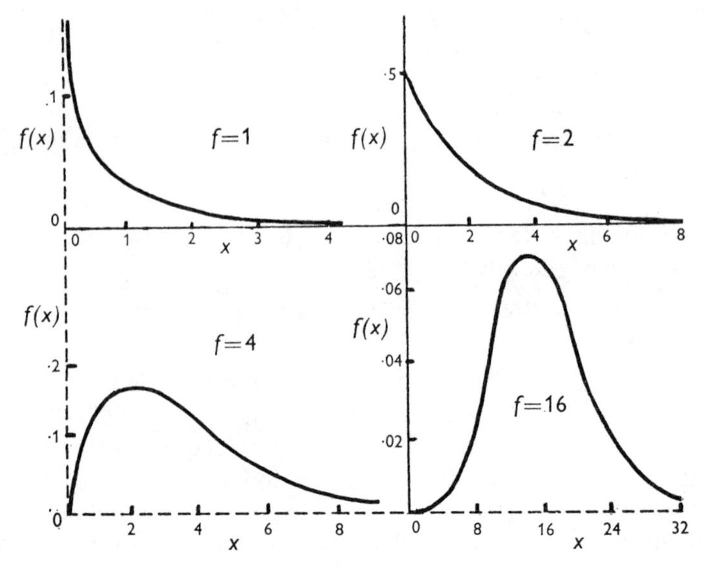

FIG. 24. The density function of the χ^2 distribution with 1, 2, 4 and 16 degrees of freedom

An important property of the distribution is that if Y_1 and Y_2 are independent χ^2 variates with f_1 and f_2 degrees of freedom then their sum, Y_1+Y_2, is a χ^2 variate with f_1+f_2 degrees of freedom. This follows immediately from the definition of the distribution; for if Y_1 and Y_2 are respectively the sums of the squares of f_1 and f_2 mutually independent standard normal variates then Y_1+Y_2 is the sum of the squares of f_1+f_2 such variates. This result can also be proved by considering the moment generating function of Y_1+Y_2.

Many of the practical applications of the χ^2 distribution derive from the following theorem. Suppose that $Z_1, Z_2, \ldots, Z_n$

are independent, standard normal variates and that $Y_1, Y_2, \ldots, Y_n$ are linear functions of them

$$Y_1 = a_1Z_1+a_2Z_2+\ldots+a_nZ_n$$
$$Y_2 = b_1Z_1+b_2Z_2+\ldots+b_nZ_n$$

and so on. The theorem states that the Y_i's will be independent, standard normal variates if, and only if, (1) all quantities like $\sum a_i^2$ are equal to 1, and (2) all quantities like $\sum a_ib_i$ are zero. Verbally this means that the sums of the squares of the coefficients must be unity and the sums of their cross-products zero. A linear transformation which satisfies these conditions is called *orthogonal*. The essential feature of such a transformation is that it preserves distance from the origin so that $\sum Y_i^2 = \sum Z_i^2$ identically; it may therefore be interpreted geometrically as a rotation of the axes. For a fuller discussion the reader is referred to a textbook on algebra such as Ferrar (1941).

To prove the theorem we shall for simplicity consider only two variables,

$$Y_1 = a_1Z_1+a_2Z_2$$
$$Y_2 = b_1Z_1+b_2Z_2$$

where

$$a_1^2+a_2^2 = b_1^2+b_2^2 = 1$$
$$a_1b_1+a_2b_2 = 0.$$

It is easy to show that Y_1 and Y_2 are uncorrelated, standard normal variates. They are normally distributed since they are linear functions of independent normal variates; they have zero mean since Z_1 and Z_2 have zero mean; they have unit variance since, for example,

$$V(Y_1) = a_1^2V(Z_1)+a_2^2V(Z_2) = a_1^2+a_2^2 = 1;$$

and they are uncorrelated since

$$\begin{aligned}\text{Cov}\,(Y_1, Y_2) &= E(Y_1Y_2) = E[(a_1Z_1+a_2Z_2)(b_1Z_1+b_2Z_2)] \\ &= a_1b_1E(Z_1^2)+a_2b_2E(Z_2^2)+a_1b_2E(Z_1)E(Z_2) \\ &\qquad\qquad +a_2b_1E(Z_1)E(Z_2) \\ &= a_1b_1+a_2b_2 = 0.\end{aligned}$$

To show that Y_1 and Y_2 are not only uncorrelated but independent we first observe that the joint probability density function of Z_1 and Z_2 is

$$\frac{1}{2\pi} e^{-\frac{1}{2}(z_1^2+z_2^2)}$$

and thus depends only on the distance of the point (z_1, z_2) from the origin. If we draw the contours of equal probability

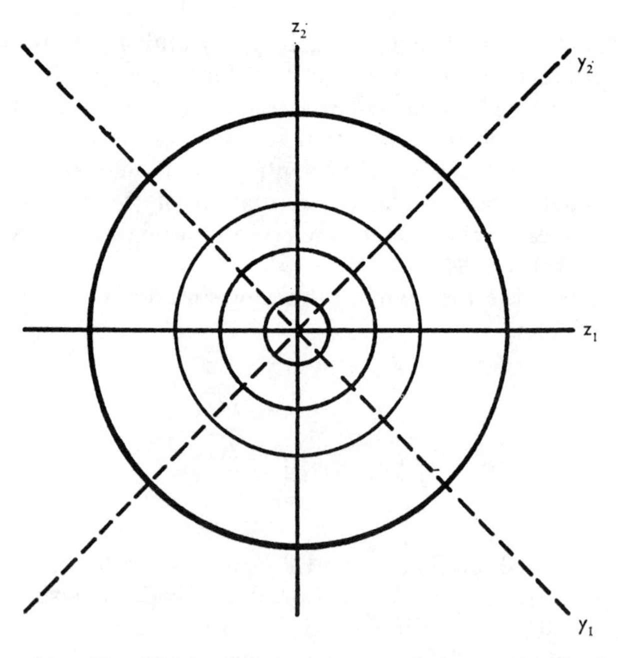

FIG. 25. The spherical symmetry of the normal distribution and its consequent invariance under rotation of the axes

density they will therefore be circles. (They would be spheres in three dimensions and hyperspheres in more than three dimensions.) It follows that if we rotate the axes about the origin the probability distribution will look the same from the new axes as it did from the old ones (see Fig. 25). But rotation of the axes corresponds algebraically to an orthogonal transformation since the essential property of such a transformation is that it preserves distances from the origin. Hence the joint probability density function of Y_1 and Y_2 is the same as that of Z_1 and Z_2; that is to say, Y_1 and Y_2 are independent, standard normal variates.

This theorem can be extended to any number of variables, although the pictorial representation cannot be interpreted literally in more than three dimensions. It is interesting to observe that Y_1 and Y_2 will be uncorrelated whatever the distribution of Z_1 and Z_2 may be, provided that they are independent and have the same variance, but that the distributions of Y_1 and Y_2 will be independent only in the normal case since this is the only distribution which has the property of spherical symmetry illustrated in Fig. 25. We shall now consider how this theorem can be applied to derive the sampling distribution of the variance.

THE SAMPLING DISTRIBUTION OF THE VARIANCE

Suppose that we have made a number of observations, $x_1, x_2, \ldots, x_n$, on some random variable and have calculated their variance, $m_2 = \sum(x_i - \bar{x})^2/n$; then, if we make n more observations, $x'_1, x'_2, \ldots, x'_n$, and calculate *their* variance, m'_2 we will in general expect it to differ slightly from m_2; and if we repeat this procedure a large number of times we shall obtain a large number of different values of the variance which will have a frequency distribution and eventually a probability distribution. This probability distribution is the sampling distribution of the variance; an observed variance may be thought of as an observation taken at random from this sampling distribution.

In considering the sampling distribution of m_2 we can as before regard the individual observations as independently and identically distributed random variables with mean μ and variance σ^2. We shall write S^2 for the sum of the squared deviations from the mean and we recall the identity

$$S^2 = \sum(x_i - \bar{x})^2 = \sum(x_i - \mu)^2 - n(\bar{x} - \mu)^2.$$

(This result follows from the formula for S^2 given on p. 56

and from the fact that S^2 is unchanged if μ is subtracted from each of the observations and from $\bar{x}$.) Now the Expected value of $\sum(x_i-\mu)^2$ is $n\sigma^2$ and the Expected value of $n(\bar{x}-\mu)^2$ is σ^2 since $E(\bar{x}) = \mu$ and $V(\bar{x}) = \sigma^2/n$. Hence

$$E(S^2) = n\sigma^2-\sigma^2 = (n-1)\sigma^2$$

and

$$E(m_2) = E(S^2/n) = E(S^2)/n = \frac{n-1}{n}\sigma^2.$$

It may seem surprising that the Expected value of the sample variance is slightly less than the population variance, σ^2. The reason is that the sum of the squared deviations of a set of observations from their mean, $\bar{x}$, is always less than the sum of the squared deviations from the population mean μ. Because of this fact S^2 is often divided by $n-1$ instead of n in order to obtain an unbiased estimate of σ^2, that is to say an estimate whose Expected value is equal to σ^2. To avoid confusion between these quantities we shall adopt the following conventional notation:

$$m_2 = S^2/n$$
$$s^2 = S^2/(n-1).$$

No assumptions have yet been made about the form of the underlying distribution, but nothing more can be said about the sampling distribution of the variance without doing so. Suppose then that the x_i's are normally distributed and consider the identity

$$\frac{\sum(x_i-\mu)^2}{\sigma^2} = \frac{S^2}{\sigma^2} + \frac{n(\bar{x}-\mu)^2}{\sigma^2}.$$

The left-hand side is a $\chi^2_{[n]}$ variate and the second term on the right-hand side is a $\chi^2_{[1]}$ variate. It is tempting to conjecture that S^2/σ^2 is a χ^2 variate with $n-1$ degrees of freedom. This must be so if the two terms on the right-hand side are independently distributed; for if it had any other distribution the products of the m.g.f.'s of the two variables on the right-hand side would not be equal to the m.g.f. of the variable on the left-hand side. This argument would, however, break down if the sample mean and variance were not independently distributed.

To establish this independence we write

$$Z_i = \frac{x_i - \mu}{\sigma} \qquad i = 1, \ldots n$$

$$Y_1 = \frac{1}{\sqrt{n}} Z_1 + \frac{1}{\sqrt{n}} Z_2 + \ldots + \frac{1}{\sqrt{n}} Z_n.$$

It can easily be verified that

$$\sum Z_i^2 = \frac{\sum (x_i - \mu)^2}{\sigma^2}$$

$$Y_1^2 = \frac{n(\bar{x} - \mu)^2}{\sigma^2}.$$

If we complete the transformation by defining linear functions $Y_2, Y_3, \ldots, Y_n$, in any way which satisfies the conditions of orthogonality, then the theorem of the last section shows that these linear functions of the Z_i's will be standard normal variates distributed independently both of each other and of Y_1. Furthermore, since an orthogonal transformation preserves distances it follows that

$$\sum_{i=1}^{n} Z_i^2 = \sum_{i=1}^{n} Y_i^2 = \sum_{i=2}^{n} Y_i^2 + Y_1^2$$

so that

$$\frac{S^2}{\sigma^2} = Y_2^2 + Y_3^2 + \ldots + Y_n^2.$$

It follows that S^2/σ^2 follows the χ^2 distribution with $n-1$ degrees of freedom and that its distribution is independent of that of $\bar{x}$.

This result was first obtained by Helmert in 1876 but was overlooked for some time. In his 1908 paper on 'The probable error of a mean' 'Student' conjectured that it was true, but was unable to prove it; the proof was supplied by Fisher by a geometrical method similar to the one used above which was described by 'Student' as "delving into the depths of hyperspace". Since he was unable to prove this result mathematically 'Student' tried to verify it empirically by writing down the heights of the 3000 criminals shown in Table 17 (p. 113) in random order and taking each consecutive set of 4 as a sample. The histogram of the 750 values of S^2/σ^2 obtained in this way is shown in Fig. 26 together with the theoretical χ^2 distribution with three degrees of freedom. The agreement is reasonably satisfactory although it is not quite as good as one might expect; 'Student' attributed this to the rather coarse grouping of heights which he used.

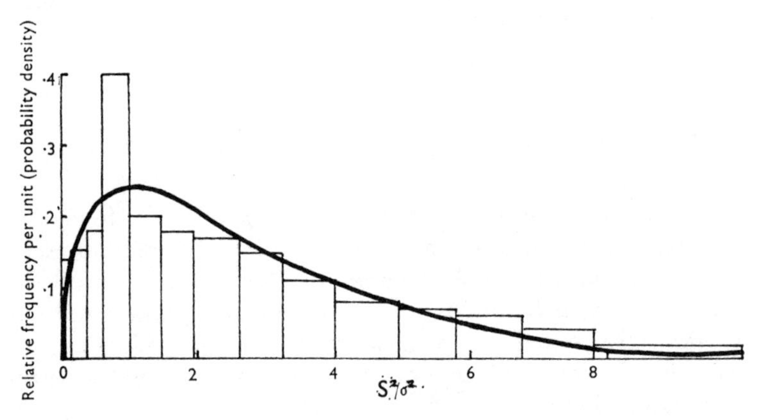

FIG. 26. Histogram of 750 values of S^2/σ^2, each based on 4 observations together with the theoretical χ^2 distribution with 3 degrees of freedom (after 'Student,' 1908)

THE *t* DISTRIBUTION

Suppose that Z is a standard normal variate and that Y independently follows a χ^2 distribution with f degrees of freedom. The random variable

$$T = \frac{Z}{\sqrt{Y/f}}$$

is said to follow the t distribution with f degrees of freedom. This distribution was first considered by 'Student' in his 1908 paper in which he discussed how one should allow for the error introduced by the fact that the standard deviation of the mean, $\sigma/\sqrt{n}$, is not known exactly but is estimated by the variable quantity, $s/\sqrt{n}$. In repeated samples of size n from a normal population with mean μ and variance σ^2 the sample mean $\bar{x}$ will be normally distributed with mean μ and variance σ^2/n and S^2/σ^2 will independently follow a χ^2 distribution with $n-1$ degrees of freedom. Hence

$$\frac{(\bar{x}-\mu)}{\sigma/\sqrt{n}} \bigg/ \sqrt{\frac{S^2}{(n-1)\sigma^2}} = \frac{(\bar{x}-\mu)}{s/\sqrt{n}}$$

will follow the t distribution with $n-1$ degrees of freedom. If we use the estimated standard deviation, s, instead of the true unknown value, σ, in making inferences about μ we must therefore employ the t distribution instead of the normal distribution. This distribution is therefore of great importance in statistical inference as will become clear in the next chapter.

It can be shown (see Problem 8.6) that the density function of T is

$$f(t) = \text{const} \times \left(1+\frac{t^2}{f}\right)^{-\frac{1}{2}(f+1)}$$

where the constant is

$$\frac{A(f+1)}{A(f)\sqrt{2\pi f}}.$$

The density function of the t distribution is shown in Fig. 27 for different values of f. When f is large it tends to the standard normal distribution since the Expected value of Y/f is 1 and its variance is $2/f$ which tends to zero as f increases; hence the denominator of T will cluster more and more closely about 1 as f becomes large. When f is small the distribution is far

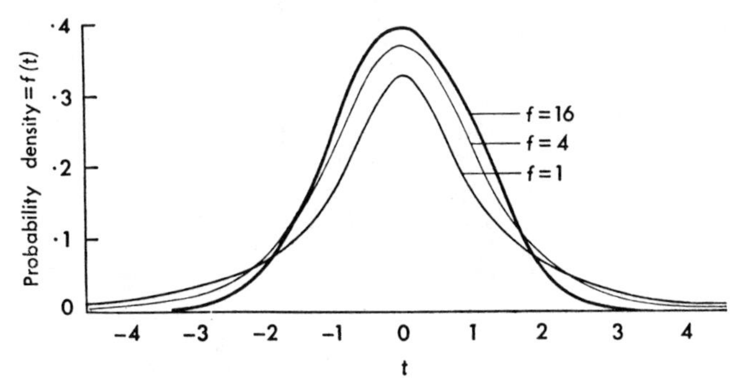

FIG. 27. The density function of the t distribution for different numbers of degrees of freedom

from normal; it is always symmetrical about zero but the variance increases as f becomes smaller owing to the increasing variability of the denominator and the shape of the distribution becomes leptokurtic. Indeed, when f is 4 the fourth moment becomes infinite, and when f is 1 or 2 the distribution is of such an extreme leptokurtic type that the variance is infinite. In practice, of course, one will never use the t distribution with so small a number of degrees of freedom since no one would hope to draw conclusions from 2 or 3 observations. The t distribution with 1 degree of freedom whose density function is

$$f(t) = \frac{1}{\pi(1+t^2)}$$

is also known as the Cauchy distribution. It is useful in constructing counter-examples to theorems such as the central limit theorem. The percentage points of the t distribution are tabulated at the end of the book. It will be seen that the normal distribution can be used instead without serious error when f is greater than 20.

We have already discussed 'Student's' empirical verification

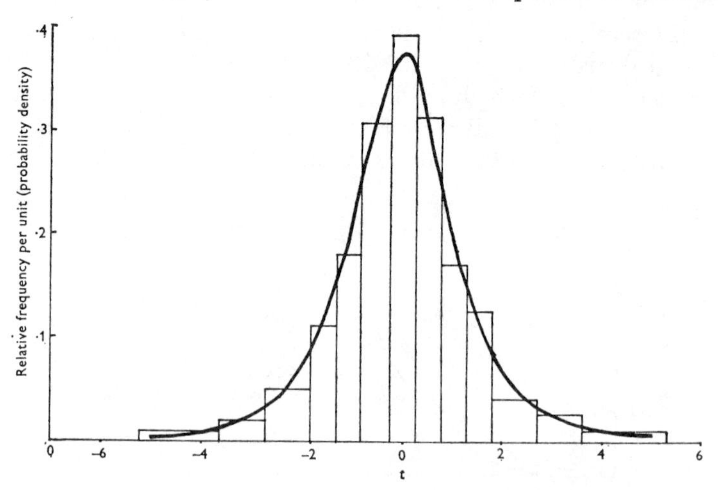

FIG. 28. Histogram of 750 values of $t = \dfrac{\bar{x}-\mu}{s/\sqrt{n}}$, each based on 4 observations together with the theoretical t distribution with 3 degrees of freedom (after 'Student', 1908)

of the χ^2 distribution as the sampling distribution of S^2/σ^2. His corresponding results for the t distribution with 3 degrees of freedom as the sampling distribution of

$$t = \frac{\bar{x}-\mu}{s/\sqrt{n}}$$

for 750 values of t each based on four observations are shown in Fig. 28. The agreement with the theoretical curve is good.

The F Distribution

Suppose that Y_1 and Y_2 are independent χ^2 variates with f_1 and f_2 degrees of freedom respectively. The random variable

$$V = \frac{Y_1/f_1}{Y_2/f_2}$$

is said to follow the F distribution with f_1 and f_2 degrees of freedom. This distribution was so named in honour of Fisher who first studied it in a rather different form in 1924. It is the sampling distribution of the ratio of two independent, unbiased estimates of the variance of a normal distribution and has widespread application in the analysis of variance.

It can be shown (see Problem 8.7) that the density function of V is

$$g(v) = \text{const} \times \frac{v^{\frac{1}{2}f_1-1}}{(f_1v+f_2)^{\frac{1}{2}(f_1+f_2)}} \qquad 0 \leqq v < \infty$$

where the constant is

$$f_1^{\frac{1}{2}f_1} f_2^{\frac{1}{2}f_2} \frac{A(f_1+f_2)}{A(f_1)A(f_2)}.$$

The mean of the distribution is approximately 1 provided that f_2 is not too small, since it is the distribution of the ratio of two random variables each of which has unit mean. The mean is not exactly 1, since the Expected value of $1/Y_2$ is only approximately $1/f_2$. The variance of the distribution decreases as f_1 and f_2 increase since the variance of Y_1/f_1 decreases with f_1 and the variance of Y_2/f_2 with f_2. The percentage points are tabulated at the end of the book.

It is interesting to consider the relationship between the χ^2, t and F distributions. If T is a t variate with f degrees of freedom then T^2 is an F variate with 1 and f degrees of freedom since the square of a standard normal variate is by definition a $\chi^2_{[1]}$ variate. Similarly, if Y is a χ^2 variate with f degrees of freedom then Y/f is an F variate with f and ∞ degrees of freedom since as f_2 tends to infinity Y_2/f_2 becomes virtually constant. The F distribution thus constitutes a wide class of distributions, depending on the parameters f_1 and f_2, which covers the χ^2 and t distributions as special cases, the former with $f_2 = \infty$ and the latter with $f_1 = 1$.

Exercises

8.1. The intelligence quotient is approximately normally distributed with $\mu = 100$ and $\sigma = 16$. In a randomly selected group of 10 people what is the probability that the observed standard deviation, s, will be less than 8·8? (P. P. P., Trinity, 1965)

8.2. If four observations are taken from a normal distribution what is the probability that the difference between the observed mean $\bar{x}$ and the true mean μ will be less than five times the observed standard deviation, s?

8.3. Suppose that the true standard deviation of technician A in performing some chemical determination is half that of technician B. If they each do 7 determinations on the same solution what is the chance that B will appear to be more accurate than A as judged by their observed standard deviations? (Assume that the errors are normally distributed.)

Problems

8.1. If Z is a standard normal variate find the distribution of $Y = Z^2$ by a direct method (cf. Problem 3.3).

8.2. If $A(f)$ is defined as on p. 125 show, by partial integration or otherwise, that $A(f-2) = (f-2)A(f)$ for $f > 2$ and hence derive the formula for $A(f)$ given on the same page. [Readers familiar with the Gamma function, should note that

$$A(f) = 2^{\frac{1}{2}f}\Gamma(\tfrac{1}{2}f).]$$

8.3. If Y_1 and Y_2 are independent χ^2 variates with f_1 and f_2 degrees of freedom prove directly that Y_1+Y_2 is a χ^2 variate with f_1+f_2 degrees of freedom (cf. Problem 3.5). Derive this result in a different way by using moment generating functions.

8.4. Investigate algebraically the sampling variance of $S^2=\Sigma(x_i-\bar{x})^2$ and show that it is $2(n-1)\sigma^4$ when the underlying distribution is normal.

8.5. Investigate algebraically the covariance of S^2 and $\bar{x}$ and show that it is zero when the underlying distribution is symmetrical.

8.6. Suppose that Z is a standard normal variate and that Y independently follows a χ^2 distribution with f degrees of freedom. Find the distribution of $V=\sqrt{Y/f}$ (cf. Problem 3.1) and hence find the density function of $T=Z/V$ (cf. Problem 3.6), which is by definition the t distribution with f degrees of freedom.

8.7. Suppose that Y_1 and Y_2 are independent χ^2 variates with f_1 and f_2 degrees of freedom respectively. Find the distribution of

$$V=\frac{Y_1/f_1}{Y_2/f_2}$$

in the same way. This is by definition the F distribution with f_1 and f_2 degrees of freedom.

8.8. A continuous random variable taking values between 0 and 1 is said to follow the Beta distribution with parameters p and q if its density function is

$$f(x)=\frac{x^{p-1}(1-x)^{q-1}}{B(p,q)} \qquad 0\leqslant x\leqslant 1$$

where $B(p, q)$, the integral of the numerator between 0 and 1, is the complete Beta function. Show that, if Y_1 and Y_2 are independent χ^2 variates with f_1 and f_2 degrees of freedom respectively, $Y_1/(Y_1+Y_2)$ follows the Beta distribution with $p=\frac{1}{2}f_1$, $q=\frac{1}{2}f_2$. Hence show that

$$B(p,q)=\frac{A(2p)A(2q)}{A(2p+2q)}\left[=\frac{(p-1)!(q-1)!}{(p+q-1)!} \text{ if } p \text{ and } q \text{ are integers.}\right]$$

[Readers already familiar with the Beta and Gamma functions will recognise the identity

$$B(p,q)=\frac{\Gamma(p)\Gamma(q)}{\Gamma(p+q)}$$

of which this is an alternative proof.]

Find the mean and variance of the Beta distribution.

8.9. Suppose that in a certain population the number of accidents occurring to an individual in a fixed time follows a Poisson distribution with mean μ, but that $a\mu$, where a is some constant, varies from person to person according to a χ^2 distribution with f degrees of freedom, where f is not necessarily an integer. Show that the distribution of the number of accidents in the whole population has the probability function

$$P(x) = \binom{x+\frac{1}{2}f-1}{x}\left(\frac{a}{a+2}\right)^{\frac{1}{2}f}\left(\frac{2}{a+2}\right)^{x} \qquad x = 0, 1, 2, \ldots$$

[Note that the binomial coefficient, $\binom{n}{x}$, is defined as

$$\binom{n}{x} = \frac{n(n-1)\ldots(n-x+1)}{x!}$$

when n is not an integer; this expression is well-defined for any value of n, positive or negative, integral or fractional, provided that x is an integer.]

If we redefine its parameters this distribution can be re-written as

$$P(x) = \binom{x+n-1}{x}P^nQ^x = \binom{-n}{x}P^n(-Q)^x.$$

which is the appropriate term in the binomial expansion of $P^n(1-Q)^{-n}$. The distribution is for this reason called the *negative binomial distribution.* Find its probability generating function and hence find its mean and variance. Fit this distribution to the data of Table 15 on p. 96 by equating the observed with the theoretical mean and variance.

This distribution has many other applications.

Chapter 9

TESTS OF SIGNIFICANCE

Probability theory, with which we have been concerned so far, investigates questions like: Given that a coin is unbiased, what is the probability of obtaining 17 heads in 40 throws? We are now going to reverse our viewpoint and try to find answers to questions like: Given that 17 heads have occurred in 40 throws of a coin, what evidence does this provide about whether the coin is unbiased or not? Such questions are often answered by the construction of a test of significance. In this chapter we shall describe some significance tests which are in common use; the logic underlying these tests and other methods of statistical inference will be discussed in more detail in the next chapter. We shall begin by defining the nature of a significance test, using tests based on the binomial distribution as illustrations of the procedure.

Significance Tests Based on the Binomial Distribution

Suppose that a penny has been thrown one hundred times to determine whether or not it is biased. If 48 heads and 52 tails occurred it would clearly be quite consistent with the hypothesis that the true probability of heads is $\frac{1}{2}$. For although on this hypothesis the most likely result is 50 heads and 50 tails, in a large number of repetitions of a hundred throws we should expect small deviations from this ideal ratio to occur quite frequently; indeed we should be rather surprised if exactly 50 heads occurred since the probability of this event is only ·08. If, on the other hand, we observed 5 heads and 95 tails we should be led to conclude that the coin was probably biased; for although such a result *could* occur if the coin were unbiased, nevertheless one would expect to observe a deviation as large as this only very rarely and it seems more reasonable to explain it by supposing that the true probability of heads is less than a half.

It thus seems natural to suppose that the possible results of our experiment can be divided into two classes: (1) those which appear consistent with the hypothesis that the coin is unbiased because they show only a small deviation from the 50/50 ratio, and (2) those which lead us to reject this hypothesis in favour of some alternative hypothesis because they show a rather large deviation from this theoretical ratio. The question is: Where should we draw the line between these two classes, that is to say between a 'small' and a 'large' deviation?

The answer adopted in the modern theory of significance tests is that the line should be drawn in such a way that the probability of obtaining a result in the second class (the rejection class) if the coin is unbiased is equal to some small, pre-assigned value known as the level of significance. The level of significance is usually denoted by α and is often taken as 5 per cent or 1 per cent. Now we know that the number of heads in a hundred throws with an unbiased coin follows the binomial distribution with $P = \frac{1}{2}$, which can be approximated by a normal distribution with mean $nP = 50$ and standard deviation $\sqrt{nPQ} = 5$; hence the probability that it will deviate from 50 by more than $1{\cdot}96 \times 5 = 10$ in either direction is approximately 5 per cent. If we wish to work at the 5 per cent level of significance we must therefore put all results between 40 and 60 in the first class (the acceptance class) and all other results in the second class (the rejection class). This means that if we observed fewer than 40 or more than 60 heads in our experiment we should reject the hypothesis that the coin is unbiased; otherwise we should 'accept' the hypothesis that it is unbiased, that is to say we should conclude that there was no reason to suppose it biased. If we wish to work at the more stringent 1 per cent level we must arrange that the probability of an observation falling in the rejection class is only 1 per cent if the coin is unbiased; to do this we must take $2{\cdot}57 \times 5 = 13$ as the critical deviation, which means that we reject the hypothesis of no bias if the number of heads is less than 37 or more than 63.

The general significance test procedure can be stated as

follows. We are about to do an experiment to test some statistical hypothesis, which we call the *null hypothesis*; in the above example the null hypothesis was that the coin was unbiased and that the number of heads would consequently follow a binomial distribution with $P = \frac{1}{2}$. We next consider all possible results of the experiment and divide them into two classes: (1) the acceptance class, and (2) the rejection class, in such a way that the probability of obtaining a result in the rejection class when the null hypothesis is true is equal to some small, pre-assigned value, α, called the significance level. We now do the experiment. If the observed result lies in the acceptance class we 'accept' the null hypothesis as a satisfactory explanation of what has occurred; if it lies in the rejection class we 'reject' the null hypothesis as unsatisfactory.

The justification of this procedure is that if we are to reject the null hypothesis with any confidence we must know that the observed result belongs to a class of results which would only occur rather infrequently if the null hypothesis were true. For example, if we found that someone was using a criterion which rejected the null hypothesis on 50 per cent of the occasions when it was in fact true, we should clearly have little confidence in what he said. The smaller we make the significance level, α, the more confidence we shall have that the null hypothesis is really false on the occasions when we reject it; the price we pay for this increased assurance is that the smaller we make α, the more difficult it becomes to disprove a hypothesis which is false.

Nothing has been said so far about how a rejection region is to be chosen from among the large number of possible rejection regions, that is to say about how the possible results of the experiment are to be placed in decreasing order of agreement with the null hypothesis. It is very often intuitively obvious how this should be done; but attempts to find general mathematical criteria for defining 'best' rejection regions have not been very successful, except in the simplest cases. One general criterion, however, can be laid down, that the significance test should be as powerful as possible, that is to say that it should have as high a probability as possible of rejecting the null hypothesis when it is false, subject of course to the fixed probability, α, of rejecting it when it is true. The difficulty is that the power of a test depends on which of the alternative hypotheses in fact holds, and a particular test may be more powerful than a second test over one range of alternative hypotheses and less powerful over another. Thus in the coin-tossing example, instead of rejecting the hypothesis that $P = \frac{1}{2}$ whenever there are more than 60 or less than 40 heads, we could decide to reject it whenever there are more than $50+1{\cdot}64\times5 = 58$ heads and to accept it otherwise. This is called a one-tailed test. Both tests have the same probability of 5 per cent of rejecting the null hypothesis when it is true, but the one-tailed test is more powerful than the two-tailed test when P is greater than $\frac{1}{2}$ and less powerful when P is less than $\frac{1}{2}$. This is illustrated in Fig. 29. If we know beforehand that P is either equal to or greater than $\frac{1}{2}$, then the one-tailed test is preferable; but if we have no such prior knowledge then most people would choose the two-tailed test since they would want to be able to detect deviations in either direction from the null hypothesis.

For example, if 100 subjects who suffered from headaches were given two drugs and asked to report after a month's trial

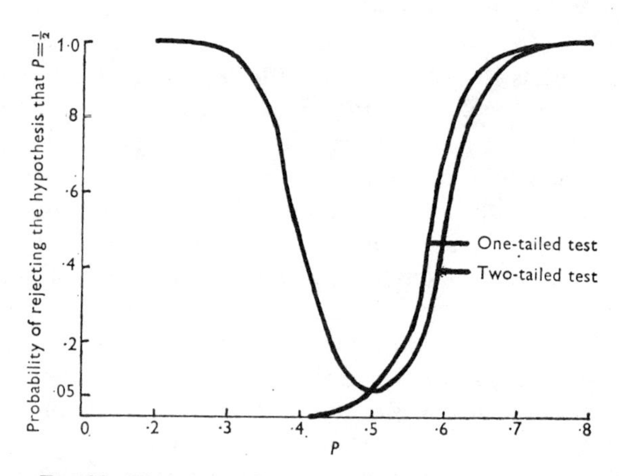

FIG. 29. The power of the one-tailed and the two-tailed tests for $P = \frac{1}{2}$ based on 100 observations with a significance level of ·05

which gave them more relief, then if the first drug was aspirin and the second a placebo we should know that the aspirin was at least as good as the placebo and we would not believe the result if the placebo appeared to be better; but if the two drugs were two different forms of aspirin we would in general have no prior knowledge about their respective merits and would choose the two-tailed test. In choosing a rejection region we must therefore consider the range of alternative hypotheses which may be true if the null hypothesis is false; a result will be classed as 'unfavourable' to the null hypothesis only in so far as it can be explained better by some alternative hypothesis. A one-sided range of alternative hypotheses gives rise naturally to a one-tailed significance test and a two-sided range of alternative hypotheses to a two-tailed test. It should be stressed, however, that the exact test to be used must be decided *before* the experiment has been done. If, in the coin-tossing example, we were to look at the result of the experiment first and decide to do a one-tailed test for $P>\frac{1}{2}$ whenever there were more than 50 heads and a one-tailed test for $P<\frac{1}{2}$ whenever there were less than 50 heads, the probability of rejecting the null hypothesis when $P=\frac{1}{2}$ would be not 5 per cent but 10 per cent; this would clearly be cheating.

It is often convenient to construct a significance test by first ordering the possible results of the experiment according to their degree of 'agreement' with the null hypothesis and then calculating the probability of obtaining a result as 'bad' as or 'worse' than the observed result; if this probability is smaller than the pre-assigned significance level, α, we reject the null hypothesis; otherwise we accept it. For example, if 38 heads have been observed in 100 throws of a coin, we calculate the probability of obtaining less than 38 or more than 62 heads if the coin is unbiased; application of the usual normal approximation shows that this probability is about ·0164. This probability is less than ·05 but greater than ·01 and we can therefore reject the hypothesis that the coin is unbiased if we are working at the 5 per cent level but not if we are working at the more stringent 1 per cent level. This procedure is equivalent to the original method of constructing a significance test and it has the advantage of determining the exact level of significance at which the null hypothesis can be rejected.

We have considered how to test the null hypothesis that the probability of success is $\frac{1}{2}$. Other values of P can be tested in the same way. Thus we find from Table 5 on p. 22 that in one of his experiments Mendel observed 5474 round seeds among 7324 seeds. On the null hypothesis that $P = \frac{3}{4}$, derived from his theory of inheritance, the number of round seeds can be regarded as a single observation from a binomial distribution with this probability, that is to say from a very nearly normal distribution with mean $nP = 5493$ and standard deviation $\sqrt{nPQ} = 37$. To test this hypothesis we therefore calculate the standardised quantity

$$d = \frac{5474-5493}{37} = -\cdot 51$$

and then find from tables of the standard normal distribution that the probability of obtaining a value as large as, or larger than, this *in absolute value* (i.e. less than $-\cdot 51$ or greater than $+\cdot 51$) is ·617. There is therefore no reason to reject the hypothesis that $P = \frac{3}{4}$. A two-tailed test is appropriate since, if P is not $\frac{3}{4}$, there is no prior reason to suppose that it will be smaller, or greater, than this value.

Instead of considering the deviation of the observed from the Expected number of successes we might equally well consider the deviation of the proportion of successes, p, from the theoretical probability, P; if P is the true value of the probability then $p-P$ will be approximately normally distributed with zero mean and variance PQ/n. The resulting significance test would be identical. This procedure can be extended to test the difference between two proportions. Suppose that two samples, of size n_1 and n_2 respectively, have been taken and that the numbers of successes in these two samples are x_1 and x_2 with corresponding proportions p_1 and p_2. If we want to test whether the probability of success is the same for the two samples we can argue that, if this is true and if this common probability is P, then p_1-p_2 should be approximately normally distributed with zero mean and variance $PQ\left(\frac{1}{n_1}+\frac{1}{n_2}\right)$ since p_1 has variance PQ/n_1 and p_2 has variance PQ/n_2. P is of course unknown, but it is unlikely that it will be very different from the overall proportion of successes in the two samples, $p = x/n$, where $x = x_1+x_2$ and $n = n_1+n_2$. We can therefore construct a significance test by calculating the quantity

$$d = \frac{p_1-p_2}{\sqrt{pq\left(\frac{1}{n_1}+\frac{1}{n_2}\right)}}$$

which should, on the null hypothesis that the two probabilities are equal, be approximately a standard normal variate.

For example, we find from Table 2 on p. 13 that the proportion of stillbirths among 368,490 male births in 1956 was ·02336 while the proportion of stillbirths among 384,250 female births was ·02239. The proportion of stillbirths in both sexes was ·02289. Hence the estimated variance of p_1-p_2 is

$$\cdot 02289 \times \cdot 97711\left(\frac{1}{368{,}490}+\frac{1}{348{,}250}\right) = 12\cdot 49 \times 10^{-8}$$

and so

$$d = \frac{\cdot 02336 - \cdot 02239}{\sqrt{12\cdot 49 \times 10^{-8}}} = 2\cdot 75.$$

The probability that a standard normal variate will exceed 2·75 in absolute value is only ·006. It can be concluded that the sex difference in the stillbirth rates is almost certainly real.

TESTS BASED ON THE t DISTRIBUTION

Suppose that we have made a number of observations, $x_1, x_2, \ldots, x_n$ on some random variable which has mean μ and variance σ^2, and that we wish to test the hypothesis that μ has some particular value which we will denote by μ_0. The natural way of doing this is to calculate the sample mean, $\bar{x}$, and see whether it is approximately equal to μ_0. If $\mu = \mu_0$ then $\bar{x}$ will, in repeated samples of size n, be approximately normally distributed with mean μ_0 and variance σ^2/n and so, if we calculate the quantity

$$d = \frac{\bar{x}-\mu_0}{\sigma/\sqrt{n}}$$

and reject the hypothesis that $\mu = \mu_0$ whenever $|d|$ is greater than 1·96, the probability of rejecting this hypothesis when it is true will be approximately 5 per cent; if we wanted to work at the 1 per cent level we would reject the hypothesis when $|d|$ is greater than 2·58, and so on. If we knew that μ cannot be smaller than μ_0 then we would use a one-tailed test and we could reject the null hypothesis at the 5 per cent level whenever d was greater than 1·64, but of course we would then be debarred from rejecting it when d was negative, however large its absolute value was.

For example, we have seen that the intelligence quotient is defined in such a way that its mean value in the general population is 100 with standard deviation 16. Suppose that we measured the I.Q. of 25 university lecturers to see whether their intelligence was above average. Our null hypothesis is that the I.Q. of university lecturers as a class has the same distribution as that of the general population. If this hypothesis were true then the average I.Q. of a group of 25 lecturers should be very nearly normally distributed with mean 100 and standard deviation $16/5 = 3{\cdot}2$; that is to say, if we computed the average I.Q.'s of a large number of randomly chosen groups of 25 lecturers we should generate a normal distribution with these parameters. We should probably be prepared to assume beforehand that the true mean I.Q. of university lecturers is not less than 100, and so we could reject the null hypothesis at the 5 per cent level if $\bar{x}$ was greater than $100+1{\cdot}64\times3{\cdot}2 = 105{\cdot}25$, since the probability that $\bar{x}$ would be greater than this value if the null hypothesis were true is 5 per cent.

The above example is rather exceptional since both the mean and the variance of the underlying distribution are specified by the null hypothesis; usually only the mean will be specified and the variance must be estimated from the data. Suppose for example that we have developed a method for estimating the concentration of some chemical compound in solution, and that in order to test the adequacy of the method we have made n estimates on a solution of known concentration μ_0. The x_i's must of course be regarded as random observations from a probability distribution with mean μ and variance σ^2, and the null hypothesis which we want to test is that $\mu = \mu_0$. How can this be done if σ^2 is unknown?

One answer to this question is that we should compute s^2, the unbiased estimator of σ^2, from the data and assume that $\sigma = s$. The argument is that if the number of observations is not too small s will be a fairly accurate estimate of σ and so we shall not go far wrong in assuming that they are the same. However, this argument breaks down if n is small. What is to be done then?

This was the question which 'Student' set out to answer in 1908 in his paper on "The probable error of a mean". Until then it had been assumed that one would always be dealing with large samples, and could use the appropriate large sample approximations, because small samples were too unreliable to be of any value. The climate of opinion at that time is illustrated by a playful remark of Karl Pearson, in a letter to 'Student' in 1912, that it made little difference whether the sum of squares was divided by n or $n-1$ in estimating the standard deviation "because only naughty brewers take n so small that the difference is not of the order of the probable error!". 'Student' had found, however, that in his practical work for Guinness' brewery he was often forced to deal with samples far too small for the customary large sample approximations to be applicable. It was gradually realised after the publication of his paper, and of R. A. Fisher's papers on other problems in small sample theory, that if the sample size were large enough the answer to any question one might ask would be obvious, and that it was only in the case of small and moderate-sized samples that any statistical problem arose. Small sample theory today lies at the heart of statistical inference.

'Student' himself stated the problem in the following words:

> Any series of experiments is only of value in so far as it enables us to form a judgment as to the statistical constants of the population to which the experiments belong. In a greater number of cases the question finally turns on the value of a mean, either directly, or as the mean difference between two quantities.
>
> If the number of experiments be very large, we may have precise information as to the value of the mean, but if our sample be small, we have two sources of uncertainty: (1) owing to the 'error of random sampling' the mean of our series of experiments deviates more or less widely from the mean of the population, and (2) the sample is not sufficiently large to determine what is the law of distribution of individuals. It is usual, however, to assume a normal distribution, because, in a very large number of cases, this gives a [very close] approximation. . . . This assumption is accordingly made in the present paper. . . . We are concerned here solely with the first of these two sources of uncertainty.
>
> The usual method of determining the probability that the mean of the population lies within a given distance of the mean of the sample is to assume a normal distribution about the mean of the sample with a standard deviation equal to $s/\sqrt{n}$, where s is the standard deviation of the sample, and to use the tables of the probability integral.
>
> But, as we decrease the number of experiments, the value of the standard deviation found from the sample of experiments becomes itself subject to an increasing error, until judgments reached in this way become altogether misleading. . . .
>
> Although it is well known that the method of using the normal curve is only trustworthy when the sample is 'large', no one has yet told us very clearly where the limit between 'large' and 'small' samples is to be drawn.
>
> The aim of the present paper is to determine the point at which we may use the tables of the probability integral in judging of the significance of the mean of a series of experiments, and to furnish alternative tables for use when the number of experiments is few.

The mathematical problem of finding the sampling distribution of

$$t = \frac{\bar{x} - \mu}{s/\sqrt{n}}$$

on the assumption that the observations come from a normal distribution with mean μ was considered in the last chapter and it was shown that this quantity should follow the t distribution with $n-1$ degrees of freedom. As an illustration of the use of the t distribution 'Student' considered some data showing the effects of two drugs in producing sleep. The sleep of ten patients was measured without hypnotic and after treatment (1) with hyoscyamine and (2) with hyoscine. The results are given in Table 19. Three questions can be asked about

TABLE 19

Additional hours' sleep gained by the use of two drugs ('Student', 1908)

Patient	Hyoscyamine	Hyoscine	Difference
1	+0·7	+1·9	+1·2
2	−1·6	+0·8	+2·4
3	−0·2	+1·1	+1·3
4	−1·2	+0·1	+1·3
5	−0·1	−0·1	0
6	+3·4	+4·4	+1·0
7	+3·7	+5·5	+1·8
8	+0·8	+1·6	+0·8
9	0	+4·6	+4·6
10	+2·0	+3·4	+1·4
$\bar{x}$	+0·75	+2·33	+1·58
s	1·79	2·00	1·23
$t = \dfrac{\bar{x}}{s/\sqrt{10}}$	+1·32	+3·68	+4·06

these data: (1) Is the first drug a soporific? (2) Is the second drug a soporific? (3) Is there any difference between the soporific effects of the two drugs? The null hypothesis in each case is that $\mu = 0$. To answer the first question we consider the first t value, $t = 1{\cdot}32$. If the first drug is not a soporific then this quantity should follow the t distribution with 9 degrees of freedom. We could presumably assume before doing the experiment that neither of the drugs would actually keep the patients awake and so a one-tailed test is appropriate. We find from tables that the probability that a t variate with 9 degrees of freedom will be greater than 1·32 is about 11 per cent; therefore we cannot reject with any confidence the hypothesis that the first drug is ineffective. On the other hand, the probability that such a t variate will be greater than 3·68 is only about 1 in 400 and we can reject with considerable confidence the hypothesis that the second drug is not a soporific.

The reader might think that these two tests taken together have answered the third question. This, however, is not so since the first test does not show that the first drug is not a soporific but only that there is insufficient evidence to justify an assertion that it is. To answer the third question we must, therefore, do another test on the ten differences shown in the last column of Table 19. The appropriate t value is 4·06. This quantity should, if the two drugs have the same soporific effect, follow the t distribution with 9 degrees of freedom. We must, however, do a two-tailed test in this case since there was no reason before the experiment was done to know which of the drugs would be more effective. The probability that a t variate with 9 degrees of freedom will be greater than 4·06 is ·0014, and so from the symmetry of the distribution the two-tailed probability is ·0028, or about 1 in 350. We can, therefore, reject with some confidence the hypothesis that the two drugs are equally effective.

'Student' considered only the problem of testing whether a set of observations with mean $\bar{x}$ could have come from a distribution with mean μ. In 1925, Fisher pointed out that the t distribution could be applied to a number of other problems, and in particular to testing whether the means of two populations are the same. Suppose that we have made m observations, $x_1, x_2, \ldots, x_m$, on a random variable X and n independent observations, $y_1, y_2, \ldots, y_n$, on another random variable Y. It will be assumed that X and Y are normally distributed with the same variance, σ^2, but that their means, μ_1 and μ_2, may be different; we wish to test whether $\mu_1 = \mu_2$. We define

$$S_1^2 = \sum_{i=1}^{m} (x_i - \bar{x})^2$$

$$S_2^2 = \sum_{i=1}^{n} (y_i - \bar{y})^2$$

$$S^2 = S_1^2 + S_2^2$$

$$s^2 = S^2/(m+n-2).$$

Then $\bar{x}-\bar{y}$ will be normally distributed with mean $\mu_1-\mu_2$ and variance $\sigma^2\left(\frac{1}{m}+\frac{1}{n}\right)$, and S^2/σ^2 will follow a χ^2 distribution with $m+n-2$ degrees of freedom independently of $\bar{x}-\bar{y}$, because of the additive property of χ^2 variates. It follows that

$$t = \frac{\bar{x}-\bar{y}}{s\sqrt{\frac{1}{m}+\frac{1}{n}}}$$

will follow the t distribution with $m+n-2$ degrees of freedom if $\mu_1 = \mu_2$.

The assumption that the two distributions have the same variance is essential to this argument. If this assumption cannot be made it would seem natural to consider instead the quantity

$$t' = \frac{\bar{x}-\bar{y}}{\sqrt{\frac{S_1^2}{(m-1)m}+\frac{S_2^2}{(n-1)n}}}.$$

This quantity is approximately a standard normal deviate if m and n are large since the variance of $\bar{x}-\bar{y}$ is $\sigma_1^2/m + \sigma_2^2/n$, where σ_1^2 and σ_2^2 are the variances of X and Y, which can be estimated by $S_1^2/(m-1)$ and $S_2^2/(n-1)$ respectively. However, t' cannot be treated as a t variate when m and n are small since the square of the denominator cannot be made into a χ^2 variate by multiplying it by a constant. Fortunately it can usually be assumed in practice that $\sigma_1^2 = \sigma_2^2$ since we most often wish to test the hypothesis that $\mu_1 = \mu_2$; it is rather unlikely that the two distributions should have the same means but different variances.

To illustrate this use of the t test Fisher considered some data from part of an electro-culture experiment at Rothamsted in 1922. Eight pots, growing 3 barley plants each, were exposed to the action of a high tension discharge, while nine similar pots were enclosed in an earthed wire cage. The number of shoots in each pot were as follows:

Electrified	16, 16, 20, 16, 20, 17, 15, 21
Caged	17, 27, 18, 25, 27, 29, 27, 23, 17.

The means of the two groups are 17·6 and 23·3 respectively, while the sums of the squared deviations from their means are 38 and 184; hence

$$S^2 = 38+184 = 222$$
$$s^2 = 222/15 = 14\cdot 6$$
$$t = \frac{17\cdot 6-23\cdot 3}{\sqrt{14\cdot 6(\frac{1}{8}+\frac{1}{9})}} = -3\cdot 2.$$

The probability that a random variable following the t distribution with 15 degrees of freedom will exceed this figure in absolute value is between 0·5 per cent and 1 per cent and we may therefore with some confidence reject the hypothesis that the means of the two populations are the same.

It should be noticed that this method is valid only if the two samples are independent of each other. It would, for instance, be quite wrong to use it on the data in Table 19 (p. 149) instead of analysing the differences in the last column; for the two observations on the same patient are quite highly correlated. This correlation is partly due to the fact that both figures are differences from the *same* control period of sleep; it is also likely that the responses of the same patient to the two drugs will be more alike than those of different patients. It would be appropriate to analyse the difference between the means rather than the mean of the differences only if the two drugs had been used on two different groups of patients.

The t test is based on the assumption that the underlying distribution is normal. When the number of observations is large we are protected by the central limit theorem since the sample mean will become approximately normally distributed almost regardless of the underlying distribution and will also be the dominating term; t will thus become approximately a standard normal variate in large samples whatever the underlying distribution may be. Several investigations, both empirical and theoretical, have been made to determine the effect of non-normality on the distribution of t in small samples. The general conclusion is that the distribution is remarkably insensitive to small or moderate departures from normality, and that one need not worry about this provided either that the non-normality is not extreme or that the number of observations is not too small; the term 'robustness' has been coined to describe this satisfactory property of a significance test. However, if neither of the above conditions is fulfilled one may be led into serious error. For example, Nair (1941) has investigated empirically the distribution of t in samples of size 6 when the underlying distribution is exponential and has found that about 10 per cent of the values were greater in absolute size than 2·57, which is the two-tailed 5 per cent point of the t distribution with 5 degrees of freedom; this means that if we used a two-tailed t test in these circumstances we should actually be working at a 10 per cent significance level when the nominal level was 5 per cent. This example is rather extreme but it shows that it is necessary to have some alternative test which can be used when the assumption of normality is not justified.

A considerable amount of work has been done in recent years on the development of non-parametric or distribution-free methods which do not necessitate any assumptions about the form of the underlying distribution. There is space here to describe only the simplest of these tests, which depend on the use of the median rather than the mean. Let us consider as an example the data on the soporific effects of hyoscine. If this drug had no effect an observation should have an equal probability of being positive or negative and the number of positive observations should, therefore, follow the binomial distribution with $P = \frac{1}{2}$. In fact, of the 10 observations 1 is negative and 9 are positive. The probability, if the null hypothesis is true, of obtaining either 1 or 0 negative observations is $(10+1)/1024 = \cdot 011$; this is a small probability and we can therefore reject the null hypothesis. It will be noticed that the probability is not as small as the probability obtained from the t test; it is clear that some information has been wasted by considering only the sign of the observations and ignoring their magnitude. The two-tailed probability, which takes into account the possibility that 9 or 10 negative observations might have been observed, is ·022, but the one-tailed test is appropriate in this case since we can predict in advance that the probability of a negative observation is either $\frac{1}{2}$ or less than $\frac{1}{2}$.

For details of other non-parametric methods, and in particular of the Mann-Whitney U test, which is in many ways more satisfactory than the tests based on medians, the reader is referred to Siegel's book *Non-parametric Statistics.* These methods have acquired a considerable vogue, particularly among social scientists and psychologists, but in my view they are rarely necessary because of the robustness of most parametric tests, and they have the serious disadvantage of being less flexible than parametric methods, so that it is difficult to adapt them to suit the situation and it is difficult to derive confidence intervals from them.

The χ^2 Test of Goodness of Fit

A problem which frequently arises is that of testing the agreement between observation and hypothesis. The most useful measure of agreement, which can be applied whenever the observations can be grouped either naturally or artificially into a finite number of classes, is the χ^2 criterion devised by Karl Pearson (1900). Suppose then that we have made n observations which can be grouped into k classes. We will write n_i for the observed number of observations and E_i for the Expected number of observations in the ith class; thus $E_i = nP_i$ where P_i is the probability, calculated from our hypothesis, that an observation will fall in the ith class. The χ^2 criterion of goodness of fit is defined as

$$\chi^2 = \sum_{i=1}^{k} \frac{(n_i - E_i)^2}{E_i}.$$

If there is perfect agreement, then $n_i = E_i$ for all i and $\chi^2 = 0$; the worse the fit, the larger χ^2 will be.

For example, we find from Table 11 on p. 82 that the observed and Expected frequencies of the number of heads in 2000 sequences of 5 spins of a coin are:

No. of heads	0	1	2	3	4	5	Total
Observed frequency	59	316	596	633	320	76	2000
Expected frequency	62·5	312·5	625	625	312·5	62·5	2000
Difference	−3·5	+3·5	−29	+8	+7·5	+13·5	0

For these data

$$\chi^2 = \frac{3{\cdot}5^2}{62{\cdot}5} + \frac{3{\cdot}5^2}{312{\cdot}5} + \ldots + \frac{13{\cdot}5^2}{62{\cdot}5} = 4{\cdot}78.$$

In this example the theoretical probabilities, P_i, are known exactly. In most applications, however, these probabilities involve one or more unknown parameters which must be estimated from the observations before the Expected numbers can be calculated. Thus in Table 12 on p. 89 the probability of a male birth had to be estimated from the observed proportion of boys before the Expected numbers of families with different numbers of boys in them could be calculated from the binomial formula; similarly in Table 17 on p. 113 the mean and variance of the observed frequency distribution had to be calculated before a normal distribution could be fitted. The fact that one or more parameters have been estimated from the data before the Expected numbers could be obtained does not affect the way in which the χ^2 criterion is calculated; for example, for the data in Table 12

$$\chi^2 = \frac{(215-165)^2}{165} + \ldots + \frac{(342-264)^2}{264} = 92{\cdot}1.$$

It does, however, affect the sampling distribution of the criterion and in consequence the interpretation to be placed upon it once it has been calculated.

An important group of applications of the χ^2 criterion is in testing for independence in contingency tables. On p. 13 we considered the following data on the sex and viability of births in England and Wales:

	Liveborn	Stillborn	Total
Male	359,881 (360,056)	8,609 (8,434)	368,490
Female	340,454 (340,279)	7,796 (7,971)	348,250
Total	700,335	16,405	716,740

Such a table is called a 2×2 contingency table since each of the characters (sex and viability) is divided into two classes. If sex and viability were independent of each other the probability of a male livebirth would be the product of the overall probabilities of these two events, which can be estimated from the corresponding marginal proportions. Hence the Expected number of male livebirths, supposing these factors to be independent, is calculated as

$$716740 \times \frac{368490}{716740} \times \frac{700335}{716740} = 360{,}056.$$

The other Expected numbers can be calculated in a similar way and are shown in brackets. The difference between observed and Expected numbers is in each case 175 in absolute value, so that the χ^2 criterion for departure from the hypothesis of independence is

$$\chi^2 = \frac{175^2}{360056} + \frac{175^2}{340279} + \frac{175^2}{8434} + \frac{175^2}{7971} = 7{\cdot}56.$$

We have seen that if there is perfect agreement between observation and hypothesis $\chi^2 = 0$ and that the worse the agreement the larger χ^2 is. In order to interpret an observed value of this criterion we must know its sampling distribution on the assumption that the hypothesis being tested is true. We shall now show that the χ^2 criterion follows *approximately* the χ^2 distribution with $k-1-p$ degrees of freedom, where k is the number of classes into which the observations are divided and p is the number of parameters which have been *independently* estimated from the data; it is from this fact that the χ^2 distribution gets its name. We must imagine that a large number of experiments have been performed in each of which n observations, classified into k groups, have been obtained and the χ^2 criterion calculated; what is the probability distribution generated by the different values of χ^2 assuming that the hypothesis being tested is true? We consider first the case in which the probabilities, P_i, are specified completely by hypothesis and no parameters need to be estimated.

Suppose then that n observations have been made and that the probability that a particular observation will fall into the ith class is P_i. The number of observations falling in the ith class will vary from experiment to experiment and is therefore a random variable. By an extension of the argument used in deriving the binomial distribution it is quite easy to show that the probability that n_1 observations will fall into the first class, n_2 into the second class and so on, is

$$P(n_1, n_2, \ldots, n_k) = \frac{n!}{n_1!n_2!\ldots n_k!} P_1^{n_1}P_2^{n_2}\ldots P_k^{n_k}$$

where, of course

$$P_1+P_2+\ldots+P_k = 1$$
$$n_1+n_2+\ldots+n_k = n.$$

This distribution is called the multinomial distribution (see Problem 6.10). The binomial distribution is a special case of the multinomial distribution with $k = 2$.

The numbers of observations are not independent random variables but are negatively correlated; for on the occasions when the number of observations in the first class, for example, is above average we should expect the other numbers to be on the average rather low since their sum is fixed. In the case of the binomial distribution there is complete negative correlation since $n_2 = n-n_1$. The n_i's can, however, be regarded as independent Poisson variates with $\mu_i = nP_i$ subject to the restriction that their sum is n; that is to say, their distribution is the same as that which would be obtained if we took k independent Poisson variates with these means and then discarded all the occasions on which they did not add up to the fixed number n. For the joint probability distribution of these independent Poisson variates is

$$e^{-nP_1}\frac{(nP_1)^{n_1}}{n_1!}\ldots e^{-nP_k}\frac{(nP_k)^{n_k}}{n_k!} = e^{-n}n^{\Sigma n_i}\frac{P_1^{n_1}\ldots P_k^{n_k}}{n_1!\ldots n_k!}$$

and the probability that they will add up to n is

$$\frac{e^{-n}n^n}{n!}$$

since, by the additive property of Poisson variates, $\sum n_i$ is itself a Poisson variate with mean $\sum nP_i = n$. If we divide the first expression by the second to obtain the conditional probability distribution of the n_i's given that they add up to n, we get the multinomial distribution.

Suppose then that $n_1, n_2, \ldots, n_k$ are independent Poisson variates with means $nP_1, nP_2, \ldots, nP_k$; then

$$Z_i = \frac{(n_i - nP_i)}{\sqrt{nP_i}}$$

has zero mean and unit variance and is approximately normally distributed provided that nP_i is not too small. Let us now make an orthogonal transformation from the Z_i's to a new set of variables Y_i in which

$$Y_1 = \sum_{i=1}^{k} \sqrt{P_i} Z_i$$

and let us impose the restraint that $\sum_{i=1}^{k} n_i = n$ which is equivalent to $Y_1 = 0$. It follows from the theorem in the last chapter that

$$\chi^2 = \sum_{i=1}^{k} Z_i^2 = \sum_{i=2}^{k} Y_i^2$$

follows approximately the χ^2 distribution with $k-1$ degrees of freedom. For this approximation to hold it is necessary that the Expected values, nP_i, should not be too small; it has been found empirically that the approximation is satisfactory, provided that each of them is greater than 5.

So far we have considered only the case when the probabilities are specified completely by hypothesis. If one or more parameters have to be estimated from the data, this will clearly decrease the average value of χ^2 since they will be estimated to make the fit as good as possible. It can in fact be shown that, provided the parameters are estimated in a reasonable way, each independent parameter estimated is equivalent to placing an additional linear restraint on the observations. Hence, if p parameters are independently estimated, χ^2 will follow approximately the χ^2 distribution with $k-1-p$ degrees of freedom.

It should be noted that in the 2×2 contingency table considered on p. 155 only two parameters, and not four, have been *independently* estimated from the data since, once the probability of a male birth has been estimated as $368490/716740 = \cdot 5141$, it follows immediately that the probability of a female birth will be estimated as $1-\cdot 5141 = \cdot 4859$ and likewise for live-birth *v.* stillbirth; the χ^2 criterion for testing the independence of these two factors therefore follows a χ^2 distribution with $4-2-1 = 1$ degree of freedom. In the general case of an $r \times s$ contingency table in which the first character is divided into r classes and the second into s classes only $r+s-2$ marginal probabilities are independently estimated since the last relative frequency in the row margin is known once the previous $r-1$ relative frequencies have been calculated, and likewise for the column margin. Hence the χ^2 criterion of independence follows the χ^2 distribution with $rs-1-(r+s-2) = (r-1)(s-1)$ degrees of freedom.

Several experiments have been done to verify the foregoing theory. In one set of experiments Yule threw 200 beans into a revolving circular tray with 16 equal radial compartments and counted the number of beans falling into each compartment. The 16 frequencies so obtained were arranged (1) in a 4×4 table, and (2) in a 2×8 table. Then χ^2 was calculated as for testing independence in a contingency table. This experiment was repeated 100 times; the observed and theoretical distributions are shown in Table 20.

TABLE 20

Distribution of χ^2 in Yule's experiment with beans (Yule and Kendall, 1950)

	4×4		8×2	
χ^2	Observed	Expected (9 d.f.)	Observed	Expected (7 d.f.)
0-5	17	17	30	34
5-10	44	48	56	47
10-15	32	26	10	15
15-20	6	7	3	3
Over 20	1	2	1	1
Total	100	100	100	100

A less artificial realisation of the distribution of the χ^2 criterion may be found in a paper by Chamberlain and Turner (1952). These authors did duplicate white cell counts on 294 slides of blood obtained from a finger puncture; 100 white cells were counted in each count and divided into three classes, neutrophils, lymphocytes and others. A typical result is given below:

	Neutrophils	Lymphocytes	Others	Total
First count	57	21	22	100
Second count	52	28	20	100
Total	109	49	42	200

For each slide the χ^2 criterion for testing the independence of the two counts was calculated ; for the above slide $\chi^2 = 1{\cdot}32$.

TABLE 21

Distribution of χ^2 in duplicate white cell counts

χ^2	Observed	Expected
0·00-0·04	5	6
0·04-0·10	11	9
0·10-0·21	14	15
0·21-0·45	29	29
0·45-0·71	26	29
0·71-1·39	59	59
1·39-2·41	62	59
2·41-3·22	19	29
3·22-4·60	25	29
4·60-5·99	26	15
5·99-7·82	9	9
Over 7·82	9	6
Total	294	294

The distribution of the 294 values of χ^2 is shown in Table 21 together with the theoretical χ^2 distribution with 2 degrees of freedom (exponential distribution).

It remains to construct a significance test. Since small values of χ^2 indicate good agreement with hypothesis we wish to reject the hypothesis only when χ^2 is large. We therefore calculate the probability of obtaining a value of χ^2 greater than the observed value on the assumption that the hypothesis is true and that in repeated sampling the χ^2 criterion will therefore follow the χ^2 distribution with the appropriate number of degrees of freedom; if this probability is small we reject the hypothesis, otherwise we accept it.

For example, for the coin data considered on p. 154 at the beginning of this section χ^2 is 4·78 with 5 degrees of freedom; the probability, P, that a χ^2 variate with 5 degrees of freedom will be greater than 4·78 is ·44 and there is thus no reason to reject the hypothesis that the data follow a binomial distribution with probability $\frac{1}{2}$. For the sex ratio data of Table 12 considered on p. 155 χ^2 is 92·1 with 7 degrees of freedom ($k = 9, p = 1$) and P is very small (less than one in a million); we can therefore reject the hypothesis that these data follow a binomial distribution exactly with some confidence, although inspection of the table will show that the *proportional* deviations from the binomial distribution are fairly small, and that it is thus not a bad approximation. For the 2×2 contingency table on the sex and viability of births in England and Wales (p. 155) $\chi^2 = 7{\cdot}56$ with 1 degree of freedom for which $P = {\cdot}006$; it can be concluded that these factors are probably not independent. The result of this test is the same as that obtained by comparing the proportions of stillbirths in the two sexes on p. 145; the two tests are in fact identical since the χ^2 value with 1 degree of freedom is identically equal to the square of the standard normal variate considered there.

Exercises

9.1. In one of his experiments (see Table 5 on p. 22) Mendel observed 705 plants with purple flowers and 224 plants with white flowers in plants bred from a purple-flowered $\times$ white-flowered hybrid. Test the hypothesis that the probability of a purple-flowered plant is $\frac{3}{4}$.

9.2. 200 women are each given a sample of butter and a sample of margarine and asked to identify the butter; 120 of them do so correctly. Can women tell butter from margarine?

9.3. In a similar test among 200 men, 108 identify the butter correctly; is there a sex difference in taste discrimination?

9.4. To test whether it is of advantage to kiln-dry barley before sowing, eleven varieties of barley were sown (both kiln-dried and not kiln-dried); the yields, in lb. head corn per acre, are given below:

Kiln-dried	2009	1915	2011	2463	2180	1925	2122	1482	1542	1443	1535
Not kiln-dried	1903	1935	1910	2496	2108	1961	2060	1444	1612	1316	1511

Test whether there is any advantage in kiln-drying. [' Student ', 1908]

9.5. Use the data in Table 22 on p. 210 to test (*a*) whether there is an increase in comb-growth in capons receiving $\frac{1}{2}$ mg androsterone, (*b*) whether there is an increase in capons receiving 4 mg, (*c*) whether there is any difference in the increase between capons receiving 4 mg and 8 mg.

[NOTE: in doing any part of this question ignore the data not directly relevant to it, although the rest of the data could be used in estimating σ^2.]

9.6. For Weldon's dice data in Exercise 6.4, (*a*) test whether $P=\frac{1}{2}$ by comparing the total number of successes with its Expected value, (*b*) test whether the data with $P=p$ follow a binomial distribution.

9.7. Test the goodness of fit of the Poisson distribution to the data : (*a*) in Table 13 on p. 92, (*b*) in Table 15 on p. 96, (*c*) in Exercise 6.6. In calculating χ^2 remember that no class should have an Expected value less than 5.

9.8. Test the goodness of fit of the normal distribution (*a*) to the data in Table 17 on p. 113, (*b*) to the distribution of head breadth in Table 10 on p. 40 (see Exercise 7.4).

9.9. Use the data in Table 3 on p. 18 to test (*a*) whether the red die is unbiased, (*b*) whether the white die is unbiased, (*c*) whether the two dice are independent.

9.10. Consider a typical 2×2 table:

	A	not A	Total
B	a	b	$a+b$
not B	c	d	$c+d$
Total	$a+c$	$b+d$	n

Show that the χ^2 criterion for testing independence is given by the formula

$$\chi^2=\frac{n(ad-bc)^2}{(a+b)(c+d)(a+c)(b+d)}$$

9.11. Fisher quotes the following data of Lange on the frequency of criminality among the monozygotic and dizygotic twins of criminals:

	Convicted	Not convicted	Total
Monozygotic	10	3	13
Dizygotic	2	15	17
Total	12	18	30

Test whether monozygotic twins of criminals are more likely to be criminals themselves than are dizygotic twins by calculating χ^2 (*a*) from the formula, $\chi^2 = \sum \frac{(O-E)^2}{E}$, (*b*) from the formula in the previous exercise.

(In (a), O and E stand respectively for the observed and the expected number of observations in a class.)

9.12. Yates suggested that, when the numbers in a 2×2 table are small, a better approximation to the χ^2 distribution would be obtained from the criterion

$$\chi^2 = \sum \frac{(|O-E| - \frac{1}{2})^2}{E}$$

in which the absolute value of each deviation is reduced by $\frac{1}{2}$ before it is squared, on the analogy of the correction for continuity in evaluating the tail of the binomial distribution by the normal distribution (see p. 119). Show that this corrected χ^2 criterion can be calculated from the formula

$$\chi^2 = \frac{n(|ad-bc| - \frac{1}{2}n)^2}{(a+b)(c+d)(a+c)(b+d)}$$

Calculate the χ^2 criterion with Yates' correction for the data in the previous exercise (*a*) from the original definition, (*b*) by using the above formula.

Problems

9.1. A sampling scheme to control the quality of loads of bricks consisted of taking simple random samples of size n from each load, determining their mean specific gravities $\bar{x}$, and accepting a load as satisfactory or rejecting it according as $\bar{x} \leqslant c$ or $\bar{x} > c$. It was decided that if μ, the mean specific gravity of a load, exceeded 2·38 the probability of its rejection should be at least 0·99 while if $\mu \leqslant 2{\cdot}36$, the probability of its acceptance should be at least 0·95. The standard deviation of specific gravity in a load could be taken as 0·015. Find the smallest value of n satisfying these requirements and the corresponding value of c. [Certificate, 1958]

9.2. Consider the typical 2×2 table in Exercise 9.10. If the two criteria of classification were male/female and liveborn/stillborn and if these were independent of each other the observed numbers (a, b, c, d) would follow a multinomial distribution and the two sets of marginal totals would follow independent binomial distributions with probabilities depending on the probabilities of male birth and of stillbirth. On the other hand, if we were testing the effect of some drug and the criteria were cured/not cured and treated/untreated then the numbers of patients treated and untreated is fixed by the experimenter and is not a random variable; in this case the numbers of people cured in the treated and untreated groups follow independent binomial distributions, and the total number cured also follows a binomial distribution if the treatment has no effect. Write down the probabilities of the observed table and of the marginal totals in each case on the assumption of independence and hence show that in both cases the probability of obtaining the observed numbers, *conditional on the marginal totals being what they are*, is

$$\frac{(a+b)!(c+d)!(a+c)!(b+d)!}{n!a!b!c!d!}$$

Note that this is independent of the unknown probabilities of the events.

9.3. A significance test for independence in a 2×2 table can be obtained by calculating the probability of the observed table, conditional on the marginal totals being as observed, together with the probabilities of all tables with the same marginal totals which are more extreme (i.e. have a lower probability) than the observed table. This exact test, which can be used however small the numbers in the table, is due to R. A. Fisher (1925) who illustrated it by considering Lange's data on the frequency of criminality among the monozygotic and dizygotic twins of criminals (Exercise 9.11). Perform Fisher's exact test on these data using (*a*) a one-tailed, (*b*) a two-tailed test. Which test is appropriate?

9.4. If several independent significance tests have been made of a hypothesis, giving significance levels $P_1, P_2, \ldots, P_n$, the overall level of significance cannot be obtained by multiplying these probabilities together. Why not?

If X is uniformly distributed between 0 and 1 prove that $-2 \log_e X$ is a χ^2 variate with 2 d.f. (see Problem 3.2) and show how this fact can be used to combine the results of independent significance tests. If three tests have given significance levels of ·145, ·263 and ·087 how would you assess their overall significance?

Chapter 10

STATISTICAL INFERENCE

Considerable controversy has arisen about the general principles which should be adopted in making statistical inferences. Unfortunately this controversy has been marred in the past by personal polemics, in consequence of which many statisticians are reluctant to continue the debate in public. This seems to me doubly unfortunate since it is impossible to form a balanced view of statistical inference unless all sides of the argument are considered and since it is also a subject of great intrinsic interest. In this chapter I shall, therefore, try to give a critical account of these different points of view. This account will inevitably be coloured by the fact that my own sympathies lie with the adherents of the frequency viewpoint, who rely on significance tests and confidence intervals; but I can see the attractions of other viewpoints and I shall try to present their case fairly and without rancour. The reader must eventually make up his own mind. I shall begin by discussing the ' orthodox ' methods developed by the frequency school of thought.

Significance Tests and Confidence Intervals

The significance test procedure was defined in the preceding chapter; we must now consider how the result of such a test is to be interpreted. Firstly, a popular misconception must be cleared up. The rejection of a hypothesis at the 5 per cent level does not imply that the probability that the hypothesis is false is ·95; it merely implies that the observed result belongs to a class of results whose overall probability of occurrence, if the null hypothesis is true, is ·05. This provides good reason, in the sense of rational degree of belief, for supposing the hypothesis to be false, but no numerical value can be placed upon this degree of belief.

It is also important to understand the meaning of 'accepting' and 'rejecting' the null hypothesis. If a significance test results in the acceptance of the null hypothesis, it does not follow that we have grounds for supposing this hypothesis to be true but merely that we have no grounds for supposing it to be false. If only a few observations have been made, the power of the test will be low and we shall be unlikely to detect even quite large departures from the null hypothesis; failure to reject the null hypothesis under such circumstances clearly cannot be interpreted as positive evidence in its favour. If a large number of observations have been made, common sense tells us that the null hypothesis is unlikely to be far from the truth when it is accepted, since the test will be very powerful and there will be a high probability of detecting even a small departure from the null hypothesis. We require further information, however, to turn this common sense interpretation into an exact argument. Strictly speaking, therefore, a significance test is a one-edged weapon which can be used for discrediting hypotheses but not for confirming them. As Jeffreys (1961) remarks in criticising the orthodox theory of significance tests: "The scientific law is thus (apparently) made useless for purposes of inference. It is merely something set up like a coconut to stand until it is hit."

If a significance test results in the rejection of the null hypothesis we have reason to suppose that it is false, but we have no information about *how* false it may be. This leads to the conclusion that a significance test can be too powerful. For we do not usually expect any hypothesis to be *exactly* true and so we do not want to reject a hypothesis unless we can be fairly sure that it differs from the true hypothesis by an amount sufficient to matter. If, however, we take a large enough sample and the power of the test becomes sufficiently high, we may expect any hypothesis differing, however little, from the null hypothesis to be rejected. This paradox was first pointed out by Berkson (1938) in discussing the χ^2 test of the goodness of fit of a normal distribution. He writes: "We may assume that it is practically certain that any series of real observations does not follow a normal curve *with absolute exactitude* in all respects, and, no matter how small the discrepancy between the normal curve and the true curve of observations, the chi-square P will be small if the sample has a sufficiently large number of observations in it."

The significance test procedure thus suffers from two serious disadvantages, that it can only be used for discrediting hypotheses, and that we do not know whether the discredited hypothesis is a good approximation to the truth which can be retained as such or whether it is a bad approximation which must be discarded. These disadvantages are largely overcome in the method of confidence intervals which was first explicitly formulated by Neyman and Pearson in the 1930s and which will now be discussed.

A confidence interval is an interval constructed from the observations in such a way that it has a known probability, such as 95 per cent or 99 per cent, of containing some parameter in which we are interested. For example, if x heads and $n-x$ tails are observed in n throws of a coin, then the variable $p = x/n$ will be approximately normally distributed with mean P and variance PQ/n; to the same order of approximation we can use the estimated variance pq/n instead of the true but unknown variance. It follows that there is a probability of very nearly 95 per cent that the standardised variable $(p-P)\sqrt{n/pq}$ will be less than $1{\cdot}96$ in absolute value; but this is equivalent to saying that P will lie within the limits $p \pm 1{\cdot}96\sqrt{pq/n}$. If, therefore, we were to repeat the experiment a large number of times and at each repetition calculate these limits and assert that P lay within them, we would be correct in our assertion about 95 per cent of the time. We have thus found a method of constructing an interval which has a known probability of containing this parameter, that is to say a confidence interval for P.

Suppose, to take another example, that n observations have been made on a normal variate with mean μ and variance σ^2 so that the sample mean, $\bar{x}$, is normally distributed with mean μ and variance σ^2/n. If we knew σ we could obtain a 95 per cent confidence interval by asserting that μ lay between the limits $\bar{x} \pm 1{\cdot}96\sigma/\sqrt{n}$, since there is a 95 per cent probability that $\bar{x}-\mu$ will be less in absolute value than $1{\cdot}96\sigma/\sqrt{n}$, which is equivalent to the previous statement. If we do not know σ

we can use the estimated standard deviation, s, in its place, but we must then replace 1·96 by the corresponding percentage point of the t distribution. For we know that $\sqrt{n}(\bar{x}-\mu)/s$ follows the t distribution with $n-1$ degrees of freedom; there is, therefore, a 95 per cent probability that this quantity will be less in absolute value than t^*, the upper $97\frac{1}{2}$ per cent point of this t distribution. But this is the same as saying that μ will lie between the limits $\bar{x}\pm st^*/\sqrt{n}$, and so, if we assert that μ does lie between these limits, we shall have a 95 per cent probability of being correct; that is to say, if we make a habit of calculating these limits and asserting that μ lies between them we shall in the long run be correct in our assertion 95 times out of 100.

In interpreting the meaning of confidence intervals it is important to remember that it is the interval which is the 'random variable' and not the parameter. For example, when $\bar{x}\pm st^*/\sqrt{n}$ is taken as a confidence interval for μ, it is clear that $\bar{x}$ and s are random variables which vary from one experiment to the next and in consequence cause the interval to vary, both in its centre through variation in $\bar{x}$ and in its width through changes in s; μ, on the other hand, is a fixed but unknown constant which has a known probability of being contained somewhere in the interval. We must therefore think of the interval as a random interval which has a known probability of containing the fixed point μ. It follows that when $\bar{x}$ and s have been calculated in a particular experiment and have become known quantities it is no longer correct to say that μ has a probability of 95 per cent of lying in this *particular* interval; for this would be to regard μ as a random variable, whereas in fact it is a constant which either does or does not lie in the interval, although we do not know which.

There is a close relationship between confidence intervals and significance tests which illuminates the meaning of both these procedures. A significance test to examine the hypothesis that a parameter θ has the particular value θ_0 can be performed by constructing a confidence interval for θ and rejecting the hypothesis whenever θ_0 lies outside the interval and accepting it otherwise. Conversely, a confidence interval for θ can be constructed by performing significance tests on all possible values of θ and then including in the interval those values which are accepted and excluding from it those values which are rejected by these tests. A confidence interval is therefore much more informative than a single significance test on one particular value of the parameter, and can answer the questions which a significance test cannot answer, since it contains the results of all the significance tests on every possible value of θ.

Suppose, for example, that we want to know whether or not a coin is biased, and if so how large the bias is. If we test the hypothesis that the coin is unbiased by a significance test, we can never show that the coin is unbiased, nor can we obtain any information about the size of the bias when we can show that some bias probably exists. If, however, we place a confidence interval on the probability of heads, we can conclude that the bias is at most negligible when the confidence interval is narrow and includes, or nearly includes, $\frac{1}{2}$; and we can conclude that the bias is probably considerable when the interval does not lie near $\frac{1}{2}$. Only when the interval is wide, that is to say when only a small number of observations have been made, do we have insufficient information to be fairly sure about the *approximate* value of P.

A confidence interval therefore enables us to answer the questions which we wish to ask more satisfactorily than a significance test. The significance test procedure is only a second best upon which we can fall back in situations (for example, in testing goodness of fit) where there is no natural parameter on which to place a confidence interval. These two procedures exhaust the methods of statistical inference which have been developed by the ' orthodox ' statisticians who rely entirely upon the frequency concept of probability. We shall now consider methods which are advocated by those who are prepared to incorporate rational degrees of belief into their arguments.

Bayesian Methods

Thomas Bayes (1702-1761) was a Non-conformist minister in Tunbridge Wells. His " Essay towards solving a problem in the doctrine of chances ", which gave rise to a new form of statistical reasoning, was found among his papers after his death and published in the Philosophical Transactions of the Royal Society, of which he was a Fellow, in 1763. It is possible that he withheld it from publication during his lifetime because he came to have doubts about the validity of his method of reasoning.

Bayes's theorem can be expressed in the formula:

$$\text{Posterior probability} \propto \text{Prior probability} \times \text{Likelihood}.$$

The posterior probability is the probability that some hypothesis is true given certain evidence, the prior probability is the probability that the hypothesis was true before the evidence was collected, and the likelihood is the probability of obtaining the observed evidence given that the hypothesis is true.* This theorem therefore enables us to evaluate the probability that the hypothesis is true. It is clear that this is the sort of statement which we should like to make and which the orthodox significance test and confidence interval procedures do not enable us to make. The difficulty is that we cannot usually give a frequency interpretation to the prior probability of a hypothesis and that we are therefore forced, if we are to use this type of argument, to introduce the idea of rational degree of belief or inductive probability. First, however, let us see how Bayes's theorem works when the prior probabilities can be interpreted in the frequency sense.

It is known that among Europeans about one-third of twins born are identical and two-thirds non-identical. Thus if we know that a woman is about to give birth to twins the probability that they will be identical is $\frac{1}{3}$ and the probability that they will be non-identical $\frac{2}{3}$; these are the prior probabilities. (Strictly speaking we should take the woman's age into account since it has an effect on these probabilities, but this complication will be ignored.) Suppose now that the twins are born and that they are both male and both have the same blood groups, A, M and Rh^+; suppose furthermore that the father of the twins is found to have blood groups AB, MN, Rh^+ and the mother AB, M, Rh^-. This is our evidence.

* If the observations come from a continuous distribution their likelihood is defined as their joint probability density.

Now the probability that one of the children of such parents should be a boy with blood groups A, M, Rh^+ is the product of the probabilities of these four events which is $\frac{1}{2}\times\frac{1}{4}\times\frac{1}{2}\times\frac{5}{7}$ $=5/112$. (The value of $\frac{5}{7}$ for the probability that the child will be Rh^+ is based on the fact that about $\frac{3}{7}$ths of Rh^+ men are homozygous and $\frac{4}{7}$th heterozygous for the Rhesus gene; the required probability is thus $\frac{3}{7}+\frac{4}{7}\times\frac{1}{2}=\frac{5}{7}$.) Hence the probability that identical twins will both be male and have the observed blood groups is 5/112 and the probability that non-identical twins will both be male and have the observed blood groups is $(5/112)^2$. These are the likelihoods. Hence the probability that the twins are identical is proportional to $1/3\times5/112$ and the probability that they are not identical is proportional to $2/3\times(5/112)^2$. The ratio of these two probabilities is 112:10 and so, since they must add up to 1, the probability that the twins are identical is $112/122 = \cdot 92$ and the probability that they are not identical is $10/122 = \cdot 08$. These are the posterior probabilities. They mean that, if we had a large number of male twin births in which both the parents and the twins had the observed blood groups, about 92 per cent of the twins would be identical and 8 per cent non-identical.

It is clear that this sort of probability statement is much more direct and satisfactory than the inferences which can be drawn from the construction of a significance test or a confidence interval. Unfortunately the occasions on which the prior probabilities of the possible hypotheses are known exactly in a frequency sense are rare. When they are not known we must either try to guess them, in which case our posterior probabilities will be wrong when we guess wrong, or we must abandon the attempt to restrict ourselves to statistical probabilities and introduce prior probabilities as inductive probabilities expressing the degree of belief which it is reasonable to place in the hypotheses before the evidence has been collected; in this case the likelihood will still be a statistical probability but the posterior probability will be an inductive probability. It is the second course which is usually adopted by the advocates of Bayesian methods.

The problem which Bayes himself considered was the

following. Suppose that some event has an unknown probability, P, of occurring and that in n trials it has occurred x times and failed to occur $n-x$ times. What is the probability that P lies between two fixed values, a and b? Bayes first notes that the probability of the event occurring x times in n trials is the binomial probability

$$\frac{n!}{x!(n-x)!}\,P^x(1-P)^{n-x}.$$

This is the likelihood. He then remarks that if nothing was known about P before the experiment was done it is reasonable to suppose that it was equally likely to lie in any equal interval; hence the probability that P lay in a small interval of length dP was initially dP and so the joint probability that it lay in this interval and that the event occurs x times out of n is

$$\frac{n!}{x!(n-x)!}\,P^x(1-P)^{n-x}dP.$$

The posterior probability that P lies between a and b is thus proportional to the integral of this expression from a to b and is equal to

$$\frac{\int_a^b P^x(1-P)^{n-x}dP}{\int_0^1 P^x(1-P)^{n-x}dP}.$$

(The factor $n!/x!(n-x)!$ occurs in top and bottom and so cancels out.) This probability can be found from tables of the incomplete β function; it is closely related to the cumulative probability function of the F distribution. (cf. Problem 8.8).

This argument depends on the assumption of a uniform prior distribution of P; this prior distribution must be interpreted as a rational degree of belief. Bayes himself was aware of the difficulties of this approach, which have been discussed in Chapter 1. It is worth quoting from the accompanying letter which his friend Richard Price sent to the Royal Society. Price writes:

> I now send you an essay which I have found among the papers of our deceased friend Mr Bayes.... In an introduction

which he has writ to this Essay, he says that his design at first in thinking on the subject of it, was to find out a method by which we might judge concerning the probability that an event has to happen, in given circumstances, upon supposition that we know nothing concerning it but that, under the same circumstances, it has happened a certain number of times, and failed a certain other number of times. He adds, that he soon perceived that it would not be very difficult to do this, provided some rule could be found according to which we ought to estimate the chance that the probability for the happening of an event perfectly unknown, should lie between any two named degrees of probability, antecedently to any experiments made about it ; and that it appeared to him that the rule must be to suppose the chance the same that it should lie between any two equidifferent degrees ; which if it were allowed, all the rest might be easily calculated in the common method of proceeding in the doctrine of chances. Accordingly, I find among his papers a very ingenious solution of this problem in this way. But he afterwards considered that the *postulate* on which he had argued might not perhaps be looked upon by all as reasonable ; and therefore he chose to lay down in another form the proposition in which he thought the solution of the problem is contained, and in a scholium to subjoin the reasons why he thought so, rather than to take into his mathematical reasoning anything that might admit dispute. This, you will observe, is the method which he has pursued in this essay.

Many modern statisticians share Bayes's doubts about the possibility of expressing the prior probability quantitatively in situations where it does not have a direct frequency interpretation, and in consequence reject the Bayesian method as a general method of statistical inference. There is, however, a school of thought which advocates with vigour the use of Bayesian methods. Perhaps the clearest exposition of their views is to be found in Sir Harold Jeffreys's book *Theory of Probability*, which we shall now consider.

Jeffreys adopts the extreme view of maintaining that rational degree of belief is the only valid concept of probability. However, if the arguments considered in the first chapter of this book are accepted, it follows that the likelihood must have a frequency interpretation since it is the probability with which the observed event is predicted to occur by some hypothesis. I shall therefore modify Jeffreys's treatment by supposing that the likelihoods are statistical probabilities, but that the prior probabilities, and in consequence the posterior probabilities, are inductive probabilities.

Jeffreys recognises two sorts of statistical problem which lead to interval estimates and significance tests respectively, although of course the interpretation of these terms is not the same as in 'orthodox' statistical inference. His interval estimates are exactly the same as those of Bayes and depend on the assignment of a prior distribution to the parameter or parameters being estimated. This prior distribution will nearly always be a continuous distribution and the posterior probability that the parameter has exactly some specified value will in consequence be zero; non-zero posterior probabilities are only attached to intervals within which the parameter may lie. In a significance test, on the other hand, Jeffreys assigns half the prior probability to the null hypothesis and distributes the rest of the prior probability in some manner among the alternative hypotheses. He justifies this procedure by Occam's razor which implies that a simple hypothesis is *a priori* more likely than a complicated hypothesis; the null hypothesis will usually be simpler than the alternatives to it.

For example, if we want to test the hypothesis that the probability, P, of some event has the value P_0, we assign the prior probability $\frac{1}{2}$ to the hypothesis that $P = P_0$ and the probability $\frac{1}{2}dP$ to the hypothesis that P lies in a small interval of length dP not containing P_0. If the event has occurred x times and failed to occur $n-x$ times in n trials, the likelihood is the usual binomial probability and so the posterior probability that $P = P_0$ is proportional to

$$\frac{1}{2} \frac{n!}{x!(n-x)!} P_0^x(1-P_0)^{n-x}$$

and the posterior probability that P is not equal to P_0 is proportional to

$$\frac{1}{2} \frac{n!}{x!(n-x)!} \int_0^1 P^x(1-P)^{n-x}dP = \frac{1}{2} \frac{n!}{(n+1)!}.$$

The posterior probability of the null hypothesis is therefore

$$\frac{(n+1)!P_0^x(1-P_0)^{n-x}}{(n+1)!P_0^x(1-P_0)^{n-x}+x!(n-x)!}.$$

Thus if we throw a coin ten times and obtain 4 heads and 6 tails, the probability that the coin is unbiased, that is to say that $P = \frac{1}{2}$, is

$$\frac{11!(\frac{1}{2})^{10}}{11!(\frac{1}{2})^{10}+4!6!} = \cdot 69$$

and the probability that the coin is biased is ·31. It should be noticed that these statements are much stronger than the inferences from an orthodox significance test, which do not allow us to talk about the probability of the null hypothesis being true or false.

As a second example let us consider Jeffreys's treatment of the t test. If $x_1, x_2, \ldots, x_n$ are observations drawn from a normal distribution with mean μ and variance σ^2, their joint likelihood is

$$L = (2\pi)^{-\frac{1}{2}n}\sigma^{-n} \exp -\tfrac{1}{2}\sum (x_i-\mu)^2/\sigma^2.$$

Jeffreys now assigns a uniform prior distribution to μ, which can take any value between $-\infty$ and $+\infty$, and a uniform prior distribution to $\log \sigma$ which means that the prior probability that σ will fall in a small range of length $d\sigma$ is $d\sigma/\sigma$ since $d\log \sigma/d\sigma = 1/\sigma$. His justification for the latter step is that $\log \sigma$ can take any value between $-\infty$ and $+\infty$, whereas σ is restricted to positive values; it also has the advantage that σ^2 or $1/\sigma$ or any other power of σ has the same prior distribution since $\log \sigma^k = k \log \sigma$, whereas this is not true for a uniform prior distribution of σ.

The posterior probability that μ lies in a small interval $d\mu$ and that σ lies in a small interval $d\sigma$ is thus proportional to $Ld\mu d\sigma/\sigma$. The posterior probability that μ lies in the small interval $d\mu$ irrespective of the value of σ is therefore proportional to $d\mu \int_0^\infty (L/\sigma)d\sigma$ from which it follows (see Problem 10.3) that

$$\frac{(\mu - \bar{x})}{s/\sqrt{n}}$$

follows the t distribution with $n-1$ degrees of freedom. It should, however, be remembered that μ is the 'random variable' and that $\bar{x}$ and s are constants in this distribution; this is the exact reverse of the orthodox approach. The orthodox confidence intervals for μ are thus re-interpreted by Jeffreys as intervals within which μ has a known (inductive) probability of being.

In conclusion, it must be admitted that Bayesian methods are very attractive and answer the sort of question which one wants to ask. Many statisticians, however, have been unable to accept the prior distributions upon which they depend. One cannot, of course, deny that prior probabilities, in some sense, exist and that some hypotheses are *a priori* less plausible, and require stronger proof, than others; this is the reason why more stringent significance levels must be employed in experiments on extra-sensory perception than in more generally accepted scientific fields. The difficulty lies in assessing these prior probabilities numerically. Nevertheless, the Bayesian approach is becoming more widely used.

Fiducial Inference

The method of fiducial inference, which was proposed by Sir Ronald Fisher in 1930, purports to make direct probability statements about parameters without using prior distributions; it is thus an attempt to obtain the advantages of Bayesian methods without their arbitrary assumptions. Unfortunately, Fisher never gave a completely satisfactory account of his method, which seems to many statisticians to be based on a logical fallacy. Fisher was, however, a man of genius and even his mistakes are likely to prove interesting; his method is worth studying for this reason alone.

I shall begin by outlining Fisher's argument, relying mainly on an expository paper which he wrote in 1935 entitled 'The fiducial argument in statistical inference'. Fisher first remarks that if a sample of size n has been drawn from a normal distribution, then the quantity

$$\frac{(\bar{x}-\mu)}{s/\sqrt{n}}$$

follows a t distribution with $n-1$ degrees of freedom. If t_P is the upper $100P$ per cent point of the t distribution, the probability that this quantity will be less than t_P is therefore P. But the inequality

$$\frac{(\bar{x}-\mu)}{s/\sqrt{n}} \leqq t_P$$

is equivalent to the inequality

$$\mu \geqq \bar{x}-st_P/\sqrt{n}.$$

Fisher therefore argues that, for fixed values of $\bar{x}$ and s, the probability that the latter inequality will be true is also P. It will be noticed that μ is now the 'random variable' while $\bar{x}$ and s are fixed constants. By considering different values of t_P the probability that μ is greater than an assigned value can be determined; this procedure will therefore generate a probability distribution for μ, which Fisher calls the *fiducial distribution* of μ. This distribution can be used to construct a *fiducial interval* for μ, that is to say an interval within which μ has a high fiducial probability, such as ·95, of lying. This interval will be numerically identical with the corresponding confidence interval, although its interpretation is different since in the theory of confidence intervals μ is not regarded as a random variable known to have a high probability of lying in a fixed interval but as an unknown constant having a high probability of being contained in a random interval.

At the end of his 1935 paper Fisher applied the fiducial argument to the problem of comparing the means of two independent samples from normal populations whose variances cannot be assumed equal. (This problem had already been considered by Behrens and is often called the Behrens-Fisher problem.) The construction of a significance test in these circumstances presents considerable difficulties, as we saw in the last chapter. Fisher's solution is very simple. For if μ_1 and μ_2 are the means of the two populations, the fiducial distributions of these two parameters can be obtained as in the foregoing paragraph and the fiducial distribution of $\delta = \mu_1 - \mu_2$ can then be derived by the ordinary methods for finding the distribution of the difference between two independent random variables. This distribution can then be used to construct a fiducial interval for δ and to test the hypothesis that $\delta = 0$ in the same way as a significance test is obtained from a confidence interval. Tables for performing this test are given in Fisher and Yates' *Statistical Tables*.

Unfortunately this test is not a significance test in the normal sense of the word, since the probability of rejecting the hypothesis that $\delta = 0$ if this hypothesis is true, is not equal to the level of significance specified in the tables. In the extreme case in which both samples are of only two observations each, and when the means and the variances of the two populations are in fact equal, Fisher (1956, p. 96) has calculated that his 5 per cent fiducial criterion will be exceeded in less than 1 per cent of random trials. Fisher states that this circumstance caused him no surprise and that it was indeed to be expected, but it has convinced many statisticians that there is a fallacy in the fiducial argument.

To see where this fallacy may lie let us return to the original situation in which n observations have been made on a normal random variable with mean μ. What meaning can we attribute to the statement that there is a fiducial probability, P, that μ is greater than $\bar{x} - st_P/\sqrt{n}$? If $\bar{x}$ and s are regarded as random variables which take different values in different samples, this statement has a direct frequency interpretation; in repeated sampling from the same population the statement will be true in a proportion, P, of the samples. However, in a *particular* sample, for which $\bar{x}$ and s are known, the statement is either true or false, although we do not know which, and probability statements, in the statistical sense, have no meaning. If fiducial probability is meaningful it must therefore be an inductive probability.

We saw in the first chapter that it is difficult to assign numerical values to inductive probabilities. There is, however, one situation in which this can be done. Suppose that we know that a coin is unbiased, that is to say that the statistical probability of heads is $\frac{1}{2}$; if we toss the coin and cover it before looking at it there is clearly an inductive probability of $\frac{1}{2}$ that it is a head. Similarly, if we construct a 95 per cent confidence interval for μ, then we know that the probability that this random interval contains μ is ·95; there is therefore an inductive probability of ·95 that μ lies in a particular interval chosen at random from all such intervals.

To justify this argument it is essential that the interval should have been chosen at random from the *whole* population of intervals of which 95 per cent are known to contain μ. Suppose for example that we have made 9 observations and that $\bar{x} = 5{\cdot}2$ and $s = 3{\cdot}5$; then the 95 per cent confidence interval for μ is $5{\cdot}2 \pm 2{\cdot}7$, that is to say from 2·5 to 7·9. This interval is an interval chosen at random from an infinite population of intervals of which 95 per cent contain μ and there is therefore an inductive probability of ·95 that μ lies in this interval. But if we select from the infinite population the intervals from 2·5 to 7·9, there is not the slightest reason to suppose that 95 per cent of *these* intervals will contain μ. It might be, for example, that in Nature μ was always 0, in which case the interval 2·5 to 7·9 would never be right, although of course it would sometimes occur; or it might be that μ was always 4, in which case this particular interval would always be right; or it might be that μ was 0 half of the time and 4 the rest of the time, in which case this interval would be right half of the time.

If, therefore, we were to announce before doing the experiment that we intended to use the interval 2·5 to 7·9 as a 'confidence interval' for μ, but that the level of 'confidence' would depend on the actual values of $\bar{x}$ and s observed, there would not necessarily be a statistical probability of ·95 that μ would be contained in this interval on those occasions on which the level of 'confidence' was ·95 and we should not be justified in assigning an inductive probability of ·95 to the statement that μ lies in this interval on these occasions. It follows that fiducial probability distributions cannot be treated as if they were ordinary probability distributions. It seems to me that this is the error into which Fisher has fallen in compounding the fiducial distributions of μ_1 and μ_2 by the ordinary rules to find the fiducial distribution of $\delta = \mu_1 - \mu_2$.

Statistical Decision Theory

We shall finally consider a system of statistical inference which is fashionable in America but has not yet gained much favour on this side of the Atlantic. The basic tenet of this school of thought is that statistical inference is concerned not with what to believe in the face of inconclusive evidence but with what action to take under these circumstances. This idea was first put forward by the American mathematician Abraham Wald in 1939 and was expanded in 1950 in his book, *Statistical Decision Functions*. Wald's work is of considerable mathematical difficulty and I shall therefore try in this section to describe the essentials of his theory stripped of its mathematical details.

To understand statistical decision theory it is necessary to know something about the mathematical theory of games out of which it has developed. As an example let us consider the game of two-finger Morra. This game is played by two people, each of whom shows one or two fingers and simultaneously calls his guess as to the number of fingers his opponent will show. If only one player guesses correctly, he wins an amount equal to the sum of the fingers shown by himself and his opponent; otherwise the game is a draw. If by [1 2] we indicate that a player shows one finger and guesses that his opponent will show two fingers, then the game can be set out in a table as follows:

		Player B			
		[1 1]	[1 2]	[2 1]	[2 2]
Player A	[1 1]	0	2	–3	0
	[1 2]	–2	0	0	3
	[2 1]	3	0	0	–4
	[2 2]	0	–3	4	0

The figures in the body of the table represent A's gain which is of course B's loss.

The question is, how ought A to play this game? Let us first suppose that he must always play the same strategy, and that he has only to decide which of the four strategies it is to be. He would then argue as follows:

> "If I always play the strategy [1 1], then B will eventually realise that I always show one finger and guess one finger and will therefore himself always show two fingers and guess one finger; he will therefore gain 3 points each game. If I always adopt strategy [1 2] then B will eventually always play [1 1] and will gain 2 points per game. Likewise, if I always play [2 1] B will eventually gain 4 points, and if I always play [2 2] he will eventually gain 3 points per game. Hence if I must choose a single strategy for all time the best one to adopt is [1 2] since I shall then only lose 2 points per game, whereas with any of the others I should lose more."

[1 2] is called A's *minimax* strategy since it minimises his maximum loss.

Player A can, however, do better than this if he changes his strategy from game to game in order to keep B guessing. Instead of choosing from the four a single strategy to which he adheres in all subsequent games he therefore assigns a probability to each of the four strategies, and at each game chooses a strategy by an appropriate random device. In deciding what probabilities to allot to the strategies he argues, as before, that his opponent will eventually discover what probabilities he is using and will react by choosing for himself the strategy which will maximise A's loss; A must therefore select those probabilities which minimise his maximum loss. It turns out that A must use only the strategies [1 2] and [2 1], and that if we write P and $Q = 1-P$ for the probabilities assigned to these two strategies, then Q/P must be between $\frac{2}{3}$ and $\frac{3}{4}$. If A adopts this strategy, then B cannot do better than do the same, in which case neither player wins any points at any game. If Q/P were less than $\frac{2}{3}$ then B could penalise A by playing [1 1]; if Q/P were greater than $\frac{3}{4}$ B could penalise A by playing [2 2]; and if A ever played either [1 1] or [2 2] B could penalise him by playing [1 2] and [2 1] with the appropriate frequencies. For a fuller account of the theory of games the reader is referred to Williams (1954) or McKinsey (1952).

Wald's theory of statistical decision is based on the idea that science can be regarded as a similar game between Nature and the experimenter. Let us consider an example from everyday life discussed in *The Foundations of Statistics* by L. J. Savage:

> Your wife has just broken five good eggs into a bowl when you come in and volunteer to finish making the omelet. A sixth egg, which for some reason must either be used for the omelet or wasted altogether, lies unbroken beside the bowl. You must decide what to do with this unbroken egg. Perhaps it is not too great an over-simplification to say that you must decide among three acts only, namely, to break it into the bowl containing the other five, to break it into a saucer for inspection, or to throw it away without inspection. Depending on the state of the egg, each of these three acts will have some consequence of concern to you, say that indicated below:

	State of egg	
Act	Good	Rotten
Break into bowl	Six-egg omelet	No omelet, and five good eggs destroyed
Break into saucer	Six-egg omelet, and saucer to wash	Five-egg omelet, and a saucer to wash
Throw away	Five-egg omelet, and one good egg destroyed	Five-egg omelet

Before you can decide on a course of action you must assign utilities to these six consequences. This is bound to be rather arbitrary, but let us suppose that a six-egg omelet is worth 6 points and a five-egg omelet 5 points; let us also suppose that a saucer to wash is worth -1 point and that each good egg destroyed costs you 2 points because of the reproaches of your wife. The utility table is then as follows:

	Good egg	Rotten egg
Break into bowl	6	-10
Break into saucer	5	4
Throw away	3	5

If from past experience you knew the probability that the egg would be rotten it would seem reasonable to choose that act which maximised your expected utility; for example, if you knew that 10 per cent of eggs were rotten and 90 per cent good, then the expected utility of breaking it into the bowl is $\cdot 9 \times 6 - \cdot 1 \times 10 = 4\cdot 4$; that of breaking it into the saucer is $4\cdot 9$ and that of throwing it away $3\cdot 2$. You should therefore break it into the saucer.

Suppose, however, that you do not have reliable information about the frequency of rotten eggs. If you were a Bayesian you would try to assign some reasonable prior probabilities to the two states of the egg and then act as if these were the known frequency probabilities. The decision theory approach is to regard this situation as a game against Nature and to adopt the strategy which will minimise your maximum loss. Wald states his position as follows: ‘ The analogy between the decision problem and a two-person game seems to be complete, except for one point. Whereas the experimenter wishes to minimise the risk, we can hardly say that Nature wishes to maximise it. Nevertheless, since Nature's choice is unknown to the experimenter, it is perhaps not unreasonable for the experimenter to behave as if Nature wanted to maximise the risk.’ It turns out that your best strategy is never to break the egg into the bowl, to break it into a saucer two-thirds of the times, and to throw it away on the remaining occasions; Nature's best strategy is to make a third of the eggs good and two-thirds rotten. If you both adopt these strategies, then your average gain will be $4\frac{1}{3}$. If you adopt any other strategy then you cannot be sure of gaining as much as this and Nature can ensure that you make less; if Nature adopts any other strategy, you can make sure of gaining more than $4\frac{1}{3}$.

Let us now consider a very simple statistical application of the theory of games. You are told that a certain coin has a probability of either $\frac{1}{4}$ or $\frac{1}{2}$ of coming up heads; you are allowed to toss the coin twice and must then decide what the probability is. If you are right, you gain 1 point; if you are wrong, you gain no points. In this decision problem you have six possible pure strategies which are listed below together with your expected gain under the two hypotheses about the coin:

Strategy	State that $P = \frac{1}{4}$	State that $P = \frac{1}{2}$	Expected gain when $P = \frac{1}{4}$	Expected gain when $P = \frac{1}{2}$
1	Never	Always	0	1
2	if $x = 0$	if $x = 1$ or 2	9/16	3/4
3	if $x = 0$ or 1	if $x = 2$	16/16	1/4
4	Always	Never	1	0
5	if $x = 1$	if $x = 0$ or 2	6/16	1/2
6	if $x = 0$ or 2	if $x = 1$	10/16	1/2

Treating this problem as a game against Nature, we find that Nature's best strategy is to choose $P = \frac{1}{4}$ with a probability of 8/14 and $P = \frac{1}{2}$ with a probability of 6/14. Your best strategy is to use strategies 2 and 3 with relative frequencies of 11:3; that is to say, you state that $P = \frac{1}{4}$ when there are no heads, $P = \frac{1}{2}$ when there are 2 heads and when there is 1 head you employ a chance device so as to have a probability of 3/14 of stating that $P = \frac{1}{4}$ and a probability of 11/14 of stating that $P = \frac{1}{2}$. If you use this mixed strategy and Nature uses her best strategy your expected gain, which is also your probability of being correct, is 9/14. No other strategy can *ensure* you as high an expected gain as this.

These ideas can be extended to more complicated situations; for details the reader is referred to the books of Chernoff and Moses (1959), Blackwell and Girshick (1954) and Wald (1950), which are in increasing order of difficulty. There are two reasons why this approach has not found favour among British statisticians. First, although some statistical problems in industry and applied science can be naturally formulated as decision problems, it seems artificial to regard the formulation and testing of hypotheses in the natural sciences in the same way. Secondly, there seems no valid reason for regarding such a decision problem as a game against Nature, since this amounts to the belief that this world is the worst of all possible worlds. Statistical decision theory gives rise to many interesting mathematical problems but it seems to many statisticians that its mathematical interest is greater than its practical value. However, this remark could have been made in their early days of many branches of mathematics which later proved to have important applications. It remains to be seen whether this will be true of statistical decision theory.

Exercises

10.1. Use the data in Table 3 on p. 18 to find a 95 per cent confidence interval for the probability of throwing a six with the white die.

10.2. Use the data in Table 2 on p. 13 to find 95 per cent confidence limits (*a*) for the stillbirth rate in males, (*b*) for the stillbirth rate in females, (*c*) for the sex difference in the stillbirth rate. [In (*c*), estimate the variance of $p_1 - p_2$ as $\frac{p_1q_1}{n_1} + \frac{p_2q_2}{n_2}$.]

10.3. Find a 95 per cent confidence interval for the mean of the distribution of litter size in rats given in Table 8 on p. 29. [Use the normal approximation. The observed mean and variance have been calculated on p. 46 and in Exercise 4.3. respectively.]

10.4. In a haemocytometer (see p. 93) each square has sides of length ·05 mm and thickness of ·01 mm. Suppose that a suspension of cells is introduced into the haemocytometer and that counts over 20 squares give a mean value of 5·2 cells per square. Find a 95 per cent confidence interval for the number of cells per c.c. in the suspension. [Use the normal approximation and estimate the variance of the distribution of cell counts from its mean.]

10.5. Use the data in Table 19 on p. 149 to find 95 per cent confidence intervals for the average number of hours' sleep gained by the use of (*a*) hyoscyamine, (*b*) hyoscine, and (*c*) for the superiority of the latter over the former drug.

10.6. Use the same data to find a 95 per cent confidence interval for the standard deviation of the distribution of the number of hours' sleep gained by the use of hyoscine. [Use the fact that S^2/σ^2 follows a χ^2 distribution with 9 degrees of freedom to find a confidence interval for σ^2 and hence for σ. See p. 195.]

10.7. Use the data in Table 22 on p. 210 to find 95 per cent confidence intervals (*a*) for the increase in comb-growth in capons receiving 4 mg androsterone, (*b*) for the difference in comb-growth between capons receiving 4 mg and 8 mg. [Cf. Exercise 9.5. In (*b*) note that, if the variance at the two dose levels is the same, then, in the notation of p. 150,

$$\frac{\bar{x}-\bar{y}-(\mu_1-\mu_2)}{s\sqrt{\frac{1}{m}+\frac{1}{n}}}$$

follows the t distribution with $m+n-2$ degrees of freedom.]

10.8. Suppose that a purple-flowered variety of pea is crossed with a white-flowered variety, that the resulting hybrid is allowed to self-fertilise and that one of the purple-flowered plants in the second generation, which may be either *PP* or *Pp*, is again allowed to self-fertilise and produces 10 third-generation plants which all have purple flowers. What is the probability that the second-generation plant is *PP* and will therefore always breed true? [See pp. 22-25 for the genetic background to this question.]

Problems

10.1. Suppose that a and b are normally distributed, unbiased estimators of two parameters α and β with variance $c_1\sigma^2$ and $c_2\sigma^2$ respectively and with covariance $c_3\sigma^2$, where c_1, c_2 and c_3 are known constants; and suppose that s^2 is an estimate of σ^2 such that fs^2/σ^2 is distributed as $\chi^2_{[f]}$ independently of a and b. To construct an exact confidence interval for $\rho=\alpha/\beta$, find the mean and variance of $a-\rho b$ and hence find a quadratic equation whose solution will give the required interval. This method is due to Fieller; an approximate confidence interval can be found by considering the approximate formula for the mean and variance of a/b in Problem 5.7.

Use the data in Table 22 on p. 210 to find a 95 per cent confidence interval for the ratio of the comb-growth in capons receiving 8 mg androsterone compared with those receiving 1 mg, ignoring the rest of the data.

10.2. Laplace's law of succession states that, if an event has occurred m times and failed to occur $n-m$ times in n trials, then the probability that it will occur at the next trial is $(m+1)/(n+2)$. This result is obtained by supposing that P has a uniform prior distribution between 0 and 1, so that its posterior distribution given m successes in n trials is

$$f(P)=\frac{P^m(1-P)^{n-m}}{\int_0^1 P^m(1-P)^{n-m}dP}$$

It is then argued that the probability that the event will occur at the next trial is

$$\int_0^1 Pf(P)dP.$$

Fill in the mathematical detail of the argument and examine its logical structure.

10.3. In Jeffreys' treatment of the t test (p. 175) show that

$$\int_0^\infty (L/\sigma)d\sigma \propto m_2^{\frac{1}{2}(n-1)}[m_2+(\mu-\bar{x})^2]^{-\frac{1}{2}n}$$

and hence show that

$$\frac{(\mu-\bar{x})}{s/\sqrt{n}}$$

follows the t distribution with $n-1$ degrees of freedom.

Chapter 11

POINT ESTIMATION

A *point estimate* of a parameter is a single figure, based on the observations, which is intended to be as near the true value of the parameter as possible. For example, the proportion of successes in a series of trials is the obvious estimate of the true probability of success. A point estimate is of course a random variable and it is highly unlikely that it will be exactly equal to the parameter which it estimates. In order to make any use of a point estimate it is therefore necessary to know something about its sampling distribution, and in particular about its variance. This information can then be used to construct a confidence interval, that is to say a range of values which has a known probability of containing the true value of the parameter. Thus a point estimate is of little value except as a starting point from which a confidence interval can be constructed. There is, in theory, no reason why confidence intervals should not be constructed without any reference to point estimates, but in practice the simplest procedure is to find a good point estimate of a parameter and then try to build up a confidence interval around it. This fact largely explains the importance of the theory of point estimation.

It is useful to distinguish between an estimation formula, or *estimator*, and an *estimate*, which is the value taken by the estimator on a particular occasion. For example, if x stands for the number of successes in N trials, then x/N is an estimator of P, the probability of success; if in a particular experiment $x = 40$ and $N = 100$, then $\cdot 4$ is the corresponding estimate of P. An estimator is thus a random variable and an estimate is a single observation on it. A ' good ' estimator is one whose sampling distribution is concentrated as closely as possible about the parameter which it estimates. The rest of this chapter will be devoted to a discussion of what exactly this phrase means and how such estimators can be found. We shall begin by considering why the sample mean rather than the median is usually taken as an estimator of the population mean.

The Mean *v.* The Median

Suppose that n observations have been made on a random variable whose mean, μ, we wish to estimate. We shall denote the median of the distribution by $\tilde{\mu}$; and we shall write m and $\tilde{m}$ for the mean and the median of the observations. Why should we choose m rather than $\tilde{m}$ as an estimator of μ? In some samples m will be nearer μ than $\tilde{m}$ and in others the contrary will be true, but there is no way of telling which in any particular sample. We can, however, ask which of them is more likely to be close to the true value. This question therefore demands a comparison of the sampling distributions of m and $\tilde{m}$.

Let us first suppose that the underlying distribution is asymmetrical so that $\tilde{\mu} \neq \mu$. We know that $E(m) = \mu$ and $V(m) = \sigma^2/n$. Hence, provided that σ^2 is finite, $V(m)$ will tend to zero as the number of observations increases and so the sampling distribution of m will be concentrated more and more closely about μ; m is said to be a *consistent* estimator of μ. Similarly it is intuitively clear that the sampling distribution of $\tilde{m}$ will cluster more and more closely about $\tilde{\mu}$ as the sample size increases; $\tilde{m}$ is therefore a consistent estimator of $\tilde{\mu}$. However, $\tilde{m}$ is not a consistent estimator of μ since $\tilde{\mu} \neq \mu$. Any reasonable estimator of a parameter should be consistent since we expect it to become clustered more and more closely about that parameter as the number of observations increases; indeed this was how the value of the parameter was defined in Chapter 4. We conclude that $\tilde{m}$ is not an acceptable estimator of μ unless $\mu = \tilde{\mu}$.

Let us now suppose that the underlying distribution is symmetrical so that both m and $\tilde{m}$ are consistent estimators of μ. It is clear that the sampling distributions of both these estimators are symmetrical about $\mu = \tilde{\mu}$. The variance of m is σ^2/n and its distribution tends to the normal form in large samples. We shall now show that in large samples $\tilde{m}$ is nearly normally distributed with a variance approximately equal to $1/[4nf^2(\tilde{\mu})]$ where $f(\tilde{\mu})$ is the value of the probability density function at $\tilde{\mu}$. For if the number of observations is large, the density of observations (that is, the number of observations per unit of measurement) in the neighbourhood of $\tilde{\mu}$ will be approximately $nf(\tilde{\mu})$. Hence the distance between two successive, ranked observations in this neighbourhood will be approximately $1/nf(\tilde{\mu})$. Now the number of observations greater than $\tilde{\mu}$ follows the binomial distribution with $P = \frac{1}{2}$ and is therefore approximately normally distributed with mean $\frac{1}{2}n$ and variance $\frac{1}{4}n$; and when x observations are greater than $\tilde{\mu}$, the distance between $\tilde{m}$ and $\tilde{\mu}$ is nearly $(x-\frac{1}{2}n)/nf(\tilde{\mu})$. It follows that $\tilde{m}$ is approximately normally distributed with mean $\tilde{\mu}$ and variance

$$\frac{\frac{1}{4}n}{[nf(\tilde{\mu})]^2} = \frac{1}{4nf^2(\tilde{\mu})}.$$

If the underlying distribution is normal $f(\tilde{\mu}) = f(\mu) = 1/\sigma\sqrt{2\pi} = \cdot 4/\sigma$, so that the asymptotic variance of m is $1 \cdot 57\sigma^2/n$. In large samples, therefore, the sampling distributions of m and $\tilde{m}$ differ only in their variances, the variance of $\tilde{m}$ being $1 \cdot 57$ times the variance of m. The sample median is thus more variable and is therefore a worse estimator of μ than the sample mean. The ratio of the variances of m and $\tilde{m}$, multiplied by 100, is called the *efficiency* of $\tilde{m}$ compared with m. The comparative efficiency of the median is therefore

$$100 \times \frac{V(m)}{V(\tilde{m})} = 100 \times \frac{\sigma^2/n}{1 \cdot 57\sigma^2/n} = \frac{100}{1 \cdot 57} = 64 \text{ per cent.}$$

The same accuracy can be obtained by collecting 64 observations and computing their mean as by collecting 100 observations and computing their median. (This is true because both sampling variances are inversely proportional to the number of observations.) By using the sample median instead of the mean we have in effect wasted 36 per cent of the observations.

If the underlying distribution is not normal the asymptotic variance of $\tilde{m}$ will alter. If the distribution is platykurtic (flat-topped), $f(\tilde{\mu})$ will be less than $\cdot 4/\sigma$ and the variance of $\tilde{m}$ will be greater than $1{\cdot}57\sigma^2/n$; if, however, the distribution is leptokurtic (with a high peak), $f(\tilde{\mu})$ will be greater than $\cdot 4/\sigma$ and so the variance of $\tilde{m}$ will be less than $1{\cdot}57\sigma^2/n$, and may become less than σ^2/n which is the variance of m irrespective of the shape of the distribution. The median may thus be more efficient than the mean in samples from a leptokurtic distribution. The reason for this is that a leptokurtic curve has a higher proportion of extreme observations than a normal curve with the same variance, and that the mean is more affected than the median by the occurrence of an extreme observation.

As an extreme example, consider the Cauchy distribution whose density function is

$$f(x) = \frac{b}{\pi[b^2+(x-\mu)^2]} \qquad -\infty < x < \infty.$$

This is the distribution which would be obtained if a machine gun at a distance b from an infinitely long wall and opposite the point μ on the wall were fired at it in such a way that the horizontal angle of fire was equally likely to lie in any direction (see Problem 3.4). If we put $b = 1$, $\mu = 0$, it is the t distribution with 1 degree of freedom. The tail of this distribution stretches out so far that its variance is infinite. The central limit theorem does not in consequence apply, and in fact the mean of n observations from this distribution has exactly the same distribution as a single observation (see Problem 7.5); it is thus a waste of time to take more than one observation if the sample mean is going to be used as an estimator of μ. The mean is not even a consistent estimator of μ. The sample median is much better behaved, and tends in large samples to be approximately normally distributed with mean μ and variance $\pi^2b^2/4n$.

We have seen that if the underlying distribution is normal the sample mean is a more efficient estimator of μ than the median. It is natural to ask whether there exists a more efficient estimator than the mean. This question can be answered by a remarkable theorem due to Fisher which states that the variance of an unbiased estimator of a parameter θ cannot be less than

$$\frac{1}{-E\left(\frac{d^2 \log L}{d\theta^2}\right)}$$

where L is the likelihood of the observations. An *unbiased* estimator of θ is one whose Expected value is equal to θ. The *likelihood* of a set of observations is their joint probability of occurrence if they come from a discrete distribution and their joint probability density if they come from a continuous distribution. This theorem is proved in the Appendix to this chapter.

If $x_1, x_2, \ldots, x_n$ are observations from a normal distribution with mean μ and variance σ^2 their likelihood is

$$L = (2\pi\sigma^2)^{-\frac{1}{2}n} \exp -\tfrac{1}{2}\sum(x_i-\mu)^2/\sigma^2$$

so that

$$\frac{d \log L}{d\mu} = \sum(x_i-\mu)/\sigma^2 = \frac{\sum x_i}{\sigma^2} - \frac{n\mu}{\sigma^2}$$

$$\frac{d^2 \log L}{d\mu^2} = -n/\sigma^2$$

$$\frac{-E(d^2 \log L)}{d\mu^2} = n/\sigma^2.$$

Hence the variance of an unbiased estimator of μ cannot be less than σ^2/n. If we restrict our choice to unbiased estimators and use their efficiency as a measure of their 'goodness', we cannot find a better estimator than the sample mean whose variance is exactly σ^2/n.

The use of efficiency as a measure of the 'goodness' of an estimator suffers from two drawbacks. First, it can only be used for comparing unbiased estimators since the variance of an estimator, t, is the mean of the squared deviations from the Expected value of the estimator, $E(t)$; this can only be regarded as a measure of the scatter of the estimator about the parameter, θ, which it estimates if $E(t) = \theta$. We shall see in the next section that a biased estimator may be acceptable if its sampling distribution is skew. Second, the variance is not the only measure of the scatter of the distribution which could be used. If t_1 and t_2 are two estimators of a parameter θ it is possible that t_1 has a larger variance but, for example, a smaller mean deviation than t_2; under these circumstances it would clearly be arbitrary to call t_2 a ‘ better ’ estimator than t_1. This situation can only arise if the sampling distributions of the two estimators are different in shape.

In large samples most estimators become approximately normally distributed so that neither of the above objections is valid; under these circumstances the most efficient estimator, that is to say the estimator with the smallest sampling variance, is without doubt the best estimator. (It is understood that an estimator with a symmetrical sampling distribution will be unbiased.) In small samples, however, we may wish to compare estimators whose sampling distributions are skew or have different shapes, so that the concept of efficiency is of less value. We shall see how this can be done in the next section; the discussion will be centred round the problem of estimating the variance of a distribution.

Estimation of the Variance

If t is an estimator of some parameter θ, it is clear that the ‘ centre ’ of the sampling distribution of t should be at θ; otherwise, there would be no sense in saying that t estimated θ. If the sampling distribution of t is symmetrical, there is only one possible measure of its centre; for the mean, the median and the mode coincide at the centre of symmetry. But if its distribution is skew there is not a unique measure of its centre, and the choice of a particular measure is to some extent arbitrary. The choice of an unbiased estimator whose Expected value is equal to the parameter being estimated amounts to taking the mean of the distribution as a measure of its centre.

For example, let us consider the problem of estimating the variance, and let us suppose that the underlying distribution is normal so that S^2/σ^2 follows the χ^2 distribution with $n-1$ degrees of freedom. This distribution is skew to the right and has its mean at $n-1$, its mode at $n-3$ and its median approximately at $n-\frac{5}{3}$ from the empirical rule that the median is about two-thirds of the way between the mode and the mean. If we want an unbiased estimator of σ^2 we must therefore divide the sum of squares by $n-1$; but if we want the median of our estimator to be at σ^2 we must divide the sum of squares by $n-\frac{5}{3}$. The mode is not a very sensible measure of the centre of a distribution, but a case could be made for dividing S^2 by $n-3$.

If $s^2 = S^2/(n-1)$ is taken as an estimator of σ^2, then it seems natural to take s as an estimator of σ. However, s is not an unbiased estimator of σ because $E^2(s) \neq E(s^2)$ as was explained on p. 69; $E(s)$ is in fact slightly less than σ. This difficulty is not encountered if the median is used as a measure of the centre of the distribution. For if half the values of $s'^2 = S^2/(n-\frac{5}{3})$ are less than σ^2 it immediately follows that half the values of s' are less than σ, so that the median of the distribution of s' is at σ. This invariance property of median-unbiased estimators is quite a strong argument in their favour.

A case can also be made for dividing S^2 by n or by $n+1$. The ' natural ' estimator S^2/n is also, as we shall see in the next section, the maximum likelihood estimator of σ^2 and for this reason has quite a strong intuitive appeal. Another attractive method is to seek the estimator whose mean square error, that is to say $E(t-\theta)^2$, is a minimum, on the grounds that $(t-\theta)^2$ is a measure of the ' badness ' of an estimate. If we are going to take as our estimator of σ^2 some fraction of S^2, say S^2/k, where k is a constant to be determined, then the Expected value of S^2/k is $(n-1)\sigma^2/k$ and the variance of S^2/k about this Expected value is $2(n-1)\sigma^4/k^2$ since the variance of S^2/σ^2 is $2(n-1)$. The mean square error of a biased estimator t, whose Expected value is $E(t) = \theta+b$, where b is the bias, is

$$E(t-\theta)^2 = V(t)+b^2$$

which in this case is

$$E\left[\frac{S^2}{k}-\sigma^2\right]^2 = \frac{2(n-1)\sigma^4}{k^2}+\frac{(n-1-k)^2\sigma^4}{k^2}.$$

This quantity is a minimum when $k = n+1$.

There are thus at least five constants, ranging from $n-3$ to $n+1$, whose use as divisors of the sum of squares can be justified on different grounds. The construction of a point estimator for σ^2 is therefore to some extent arbitrary; but the difficulty vanishes when we construct an interval estimate for σ^2. We know that, if the underlying distribution is normal, S^2/σ^2 follows the χ^2 distribution with $n-1$ degrees of freedom, and there is thus a probability of 95 per cent that this quantity will lie between the upper and lower $2\frac{1}{2}$ per cent points of this distribution; but this is equivalent to saying that σ^2 lies between $S^2/\chi^2_{\cdot 025}$ and $S^2/\chi^2_{\cdot 975}$, and if we habitually make this assertion we shall be correct 95 per cent of the time. For example, if we have made 15 observations and find that $S^2 = 23$, then the lower and upper $2\frac{1}{2}$ per cent points of the χ^2 distribution with 14 degrees of freedom are 5·63 and 26·12, and so the 95 per cent confidence interval for σ^2 is ·88 to 4·09. It should be noticed that no decision about the appropriate divisor for S^2 has had to be made in order to construct this interval estimate; it has, however, been assumed that all the information about σ^2 provided by the sample is contained in the sum of squares.

In the t test and in the construction of confidence intervals based on the t distribution, $n-1$ is used as a divisor of the sum of squares in order to define t. It would, however, make no difference if some other divisor were used since the distribution of t would be altered by a corresponding amount. Student in fact used n as the divisor of S^2, and his tables of the t distribution are consequently slightly different from the modern ones; but the numerical answers to which they lead in any particular case are identical.

The problem of deciding what quantity should be used as a divisor of the sum of squares is thus a pseudo-problem created by a preoccupation with point estimation rather than interval estimation. The important question is: Why should an estimator of σ^2 be based on the sum of squares rather than on some other function of the observations, such as the mean deviation or the range? The answer to this question lies in the idea of sufficiency which was developed by R. A. Fisher in the 1920s.

An estimator, t, of a parameter θ is said to be a sufficient estimator if the conditional distribution of the observations given t is independent of θ. If this is so, no more information can be squeezed out of the observations once t is known and t therefore contains all the information about θ which the observations can provide. Suppose, for example, that n observations, $x_1, x_2, \ldots, x_n$, have been made on a Poisson variate with mean μ. Their joint probability of occurrence is

$$P(x_1, x_2, \ldots, x_n) = \frac{e^{-n\mu}\mu^{\Sigma x_i}}{x_1!x_2!\ldots x_n!}.$$

The sum of the observations, $\sum x_i$, is a Poisson variate with mean $n\mu$ and so its probability of occurrence is

$$P(\sum x_i) = \frac{e^{-n\mu}\mu^{\Sigma x_i}}{(\sum x_i)!}.$$

The conditional probability of the observations given their sum is

$$\frac{P(x_1, x_2, \ldots, x_n)}{P(\sum x_i)} = \frac{(\sum x_i)!}{x_1!x_2!\ldots x_n!}$$

which does not depend on μ. The sum of the observations is thus a sufficient estimator of μ; once it has been calculated, the original observations can be thrown away since they do not contain any additional information about the parameter.

If t is a sufficient estimator of θ, then so is any monotonic function of t, such as $a+bt$, where a and b are constants; we should not choose $\sum x_i$ as the point estimator of μ in the above example, but rather $\bar{x} = \sum x_i/n$, which is also sufficient. The principle of sufficiency does not produce a unique point estimator but a class of estimators which are functions of each other. The choice of a particular point estimator from this class is to some extent arbitrary. However, we should arrive at the same confidence interval for the parameter whatever choice we made. If we are interested in interval estimation rather than point estimation the principle of sufficiency answers all our problems.

We now return to the problem of estimating the variance

of a normal distribution. The joint probability density of the observations is

$$f(x_1, x_2, \ldots, x_n) = (2\pi\sigma^2)^{-\frac{1}{2}n} \exp -\tfrac{1}{2} \frac{\sum(x_i-\mu)^2}{\sigma^2}$$

$$= (2\pi)^{-\frac{1}{2}n}\sigma^{-n} \exp -\tfrac{1}{2}\left[\frac{S^2}{\sigma^2} + \frac{n(\bar{x}-\mu)^2}{\sigma^2}\right].$$

Furthermore, $\bar{x}$ is normally distributed with mean μ and variance σ^2/n and S^2/σ^2 is independently distributed as a χ^2 variate with $n-1$ degrees of freedom. The joint probability density function of $\bar{x}$ and S^2 is therefore

$$f(\bar{x}, S^2) = f_1(\bar{x}) \times f_2(S^2)$$

$$= \frac{\sqrt{n}}{\sigma\sqrt{2\pi}} \exp -\tfrac{1}{2}n \frac{(\bar{x}-\mu)^2}{\sigma^2} \times \left(\frac{S^2}{\sigma^2}\right)^{\frac{1}{2}(n-1)-1} \exp -\tfrac{1}{2}\frac{S^2}{\sigma^2}\Big/[\sigma^2 A(n-1)]$$

$$= \left[\frac{\sqrt{n}\,S^{n-3}}{\sqrt{2\pi}\,A(n-1)}\right]\sigma^{-n} \exp -\tfrac{1}{2}\left[\frac{S^2}{\sigma^2} + \frac{n(\bar{x}-\mu)^2}{\sigma^2}\right].$$

The conditional density function of the observations given $\bar{x}$ and S^2 is $f(x_1, x_2, \ldots, x_n)/f(\bar{x}, S^2)$, which does not contain μ or σ^2; that is to say, $\bar{x}$ and S^2 contain all the information provided by the observations about μ and σ^2. They are said to be *jointly sufficient* for μ and σ^2. This is the most powerful argument we have encountered for using these statistics to make inferences about the parameters of a normal distribution. It disposes finally of the suggestion that the sample median should be used instead of the mean to estimate μ or the mean deviation instead of the standard deviation to make inferences about σ, unless we are prepared to sacrifice information for ease of computation or some other reason.

The Method of Maximum Likelihood

We have considered how different estimators can be compared and what properties are desirable in an estimator, but no general method of obtaining estimators with these properties has been described. There is one very general method of estimation which leads to estimators with a number of desirable properties. This is the method of maximum likelihood, which was developed by R. A. Fisher and which arose naturally out of his investigations into efficiency and sufficiency.

The likelihood of a set of observations has already been defined in the discussion of efficiency. The likelihood will usually depend on one or more unknown parameters which we wish to estimate. The method of maximum likelihood consists in taking as our estimators those values of the parameters which maximise the likelihood of the observations.

Suppose, for example, that n observations have been made on a Poisson variate with mean μ. Their likelihood, that is to say their probability of occurrence, is

$$L = \frac{e^{-n\mu}\mu^{\Sigma x_i}}{x_1!x_2!\ldots x_n!}.$$

We could find the value of μ which maximises the likelihood by plotting L against μ and obtaining a graphical estimate of the point where it attains its highest value. However, it is easier and more accurate to obtain the solution mathematically. For this purpose it is usually best to consider the logarithm of the likelihood; since log L is an increasing function of L it will have a maximum at the same value of μ. Now

$$\log L = -n\mu+\sum x_i \log \mu-\sum \log (x_i!).$$

The position of the maximum can be found by solving the equation

$$\frac{d \log L}{d\mu} = -n+\frac{\sum x_i}{\mu} = 0.$$

The solution is

$$\hat{\mu} = \frac{\sum x_i}{n} = \bar{x}.$$

(It is conventional to denote the maximum likelihood estimator of a parameter by placing a circumflex over it; the resulting symbol is read 'μ hat'.)

For the second example, suppose that n observations have been made on a normal variate with mean μ and variance σ^2. The logarithm of the likelihood is

$$\log L = -\tfrac{1}{2}n \log 2\pi\sigma^2 - \frac{\frac{1}{2}\sum(x_i-\mu)^2}{\sigma^2}.$$

In this case the likelihood is a function of two parameters, μ and σ^2, which must be estimated simultaneously; we must thus solve the simultaneous equations

$$\frac{\partial \log L}{\partial \mu} = \frac{\sum(x_i - \mu)}{\sigma^2} = 0$$

$$\frac{\partial \log L}{\partial \sigma^2} = -\frac{n}{2\sigma^2} + \frac{\sum(x_i - \mu)^2}{2\sigma^4} = 0.$$

The solution of the first equation is

$$\hat{\mu} = \bar{x}.$$

Substituting this value in the second equation, we find

$$\hat{\sigma}^2 = \frac{\sum(x_i - \bar{x})^2}{n} = S^2/n.$$

This example shows that maximum likelihood estimators are sometimes biased.

The method of maximum likelihood is thus a very general method which can be used whenever the likelihood of the observations can be specified. The resulting equations are not always as easy to solve as in the above examples; but, if an explicit solution cannot be found, they can be solved by an iterative, trial-and-error procedure (see Problem 11.8). This method of estimation has an obvious intuitive appeal. It also has the merit of invariance, since it is clear that if $\hat{\theta}$ is the maximum likelihood estimator of θ then $f(\hat{\theta})$ is the maximum likelihood estimator of $f(\theta)$. But the two main arguments in its favour are: first that $\hat{\theta}$ is a sufficient estimator of θ if such an estimator exists, and second that in large samples $\hat{\theta}$ becomes approximately normally distributed with Expected value approaching θ and with variance approaching the minimum possible variance of any unbiased estimator. The proof of these very desirable properties is given in the Appendix to this chapter.

The asymptotic properties of maximum likelihood estimators not only show that they become fully efficient in large samples but also enable us to find their approximate sampling distributions when the exact sampling distributions are intractable.

For example, we have already seen that if n observations have been made on a Poisson variate with mean μ then $\hat{\mu} = \bar{x}$. Furthermore,

$$\frac{d \log L}{d\mu} = -n + \frac{\sum x_i}{\mu}$$

$$\frac{d^2 \log L}{d\mu^2} = -\frac{\sum x_i}{\mu^2}$$

$$-E\left(\frac{d^2 \log L}{d\mu^2}\right) = \frac{n}{\mu} \quad \text{since } E(x_i) = \mu.$$

We conclude that in large samples $\hat{\mu}$ is approximately normally distributed with mean μ and variance μ/n, and we can use this as the sampling distribution of $\hat{\mu} = \bar{x}$ provided that n is not too small. In this case it happens that the mean and variance found by this method are exact, but this will not be true in general. It is explained in the Appendix how this method can be extended when two or more parameters are estimated simultaneously.

The method of maximum likelihood therefore leads to estimators which are sufficient and, in large samples, fully efficient and whose approximate sampling distributions can be easily found. It undoubtedly deserves the important place which it enjoys in statistical theory.

Appendix

The properties of maximum likelihood estimators

We shall assume that n observations have been made on a continuous random variable whose density function, $f(x, \theta)$, contains only one unknown parameter, θ. The discrete case can be dealt with simply by substituting the probability function, $P(x, \theta)$, for $f(x, \theta)$ in the proofs. The extension to more than one parameter will be considered later. We first prove that maximum likelihood estimators are sufficient.

The likelihood of the observations is

$$L = f(x_1, \theta) \,.\, f(x_2, \theta) \ldots f(x_n, \theta).$$

If t is an estimator of θ with density function $g(t, \theta)$ and if

$h(x_1, x_2, \ldots, x_n, \theta)$ is the conditional probability of the observations given t, the likelihood can, by the law of multiplication, be written in the alternative form

$$L = g(t, \theta)h(x_1, x_2, \ldots, x_n, \theta).$$

If, furthermore, t is a sufficient estimator of θ, h will not depend on θ and so

$$\frac{dL}{d\theta} = \frac{dg(t, \theta)}{d\theta} h(x_1, \ldots, x_n).$$

The solution of the equation

$$\frac{dL}{d\theta} = 0$$

is therefore the same as that of the equation

$$\frac{dg(t, \theta)}{d\theta} = 0$$

and so $\hat{\theta}$ must be a function of t. Hence if a sufficient estimator exists, $\hat{\theta}$ is a function of that estimator and is therefore itself sufficient. It follows in a similar way that if $t_1, t_2, \ldots, t_k$ are joint sufficient estimators of the parameters $\theta_1, \theta_2, \ldots, \theta_k$, then the maximum likelihood estimators of these parameters are functions only of $t_1, \ldots, t_k$ and are thus themselves jointly sufficient.

We shall now investigate the sampling distribution of $\hat{\theta}$. To do this we shall first consider the sampling distribution of $\frac{d \log L}{d\theta}$, where it is understood that the differential is evaluated at the true value of θ. We shall in fact show (1) that the expected value of $\frac{d \log L}{d\theta}$ is 0, (2) that its variance is $-E\frac{(d^2 \log L)}{d\theta^2}$, and (3) that in large samples it is approximately normally distributed. To illustrate these results, consider the Poisson distribution for which

$$\frac{d \log L}{d\mu} = -n + \frac{\sum x_i}{\mu}.$$

$\sum x_i$ is a Poisson variate with mean $n\mu$. The expression on the right-hand side becomes approximately normal as n becomes large, its Expected value is

$$-n+\frac{n\mu}{\mu}=0$$

and its variance is

$$\frac{n\mu}{\mu^2}=\frac{n}{\mu}$$

which is equal to

$$-E\left(\frac{d^2 \log L}{d\mu^2}\right)=-E\left(\frac{-\sum x_i}{\mu^2}\right)=\frac{n}{\mu}.$$

It is clear that $\dfrac{d \log L}{d\theta}$ is a random variable since it is a function of the random variables $x_1, \ldots, x_n$. Its distribution tends to the normal distribution in large samples since it is the sum of n independent variables,

$$\frac{d \log L}{d\theta}=\sum_{i=1}^{n} \frac{d \log f(x_i, \theta)}{d\theta}.$$

To find its Expected value we note that

$$\frac{d \log L}{d\theta}=\frac{1}{L}\frac{dL}{d\theta}.$$

Hence

$$\begin{aligned} E\left(\frac{d \log L}{d\theta}\right) &= \int \ldots \int \frac{d \log L}{d\theta} L dx_1 \ldots dx_n \\ &= \int \ldots \int \frac{dL}{d\theta} dx_1 \ldots dx_n \\ &= \frac{d}{d\theta} \int \ldots \int L dx_1 \ldots dx_n. \end{aligned}$$

The integral is by definition equal to 1 and so its first derivative is 0. (In reversing the order of integration and differentiation here and elsewhere in the Appendix it is assumed that the range of integration is independent of θ.)

To find the variance of $\frac{d \log L}{d\theta}$ we note that

$$\frac{d^2 \log L}{d\theta^2} = \frac{d}{d\theta}\left(\frac{1}{L}\frac{dL}{d\theta}\right) = -\frac{1}{L^2}\left(\frac{dL}{d\theta}\right)^2 + \frac{1}{L}\frac{d^2L}{d\theta^2}$$

$$= -\left(\frac{d \log L}{d\theta}\right)^2 + \frac{1}{L}\frac{d^2L}{d\theta^2}.$$

It follows from the same argument as before that the Expected value of the second term on the right-hand side of this expression is zero, and so

$$V\left(\frac{d \log L}{d\theta}\right) = E\left(\frac{d \log L}{d\theta}\right)^2 = -E\left(\frac{d^2 \log L}{d\theta^2}\right).$$

We turn now to the sampling distribution of $\hat{\theta}$. $\frac{d \log L}{d\theta}$ evaluated at $\hat{\theta}$ rather than at the true value of θ is by definition zero. This quantity is also approximately equal to

$$\frac{d \log L}{d\theta} + (\hat{\theta} - \theta)\frac{d^2 \log L}{d\theta^2}$$

by Taylor's theorem and so

$$\hat{\theta} - \theta \doteqdot -\frac{d \log L}{d\theta}\bigg/\frac{d^2 \log L}{d\theta^2}.$$

To the same order of approximation we can substitute the Expected for the observed value of $\frac{d^2 \log L}{d\theta^2}$ in this equation since the numerator, whose Expected value is zero, is the dominating term. It follows that the Expected value of $\hat{\theta}$ is approximately θ, its Variance approximately $1/-E\left(\frac{d^2 \log L}{d\theta^2}\right)$ and that it tends to be normally distributed in large samples.

We shall now prove Fisher's theorem on the lower limit of the variance of an unbiased estimator. If t is an unbiased estimator of θ, the covariance of t and $\dfrac{d \log L}{d\theta}$ is equal to 1 since

$$\begin{aligned}\text{Cov}\left(t, \frac{d \log L}{d\theta}\right) &= E\left(t \,.\, \frac{d \log L}{d\theta}\right) = E\left(t \,.\, \frac{1}{L}\frac{dL}{d\theta}\right) \\ &= \int \ldots \int t \frac{dL}{d\theta}\, dx_1 \ldots dx_n \\ &= \frac{d}{d\theta} \int \ldots \int tL dx_1 \ldots dx_n = \frac{d}{d\theta} E(t) = 1.\end{aligned}$$

Now the square of the covariance of two random variables must be less than the product of their variances (from the Cauchy-Schwarz inequality, or from the fact that a correlation coefficient must be between $+1$ and -1), and so

$$1 \leqq V(t) \times V\left(\frac{d \log L}{d\theta}\right)$$

whence

$$V(t) \geqq \frac{1}{V\left(\dfrac{d \log L}{d\theta}\right)} = \frac{1}{-E\left(\dfrac{d^2 \log L}{d\theta^2}\right)}.$$

These results can easily be extended to the simultaneous estimation of several parameters. If the likelihood is a function of k parameters, $\theta_1, \theta_2, \ldots, \theta_k$, it can be shown by an analogous argument that the k random variables, $\dfrac{\partial \log L}{\partial \theta_1}, \dfrac{\partial \log L}{\partial \theta_2}, \ldots,$ $\dfrac{\partial \log L}{\partial \theta_k}$, all have zero Expected value, that the variance of $\dfrac{\partial \log L}{\partial \theta_1}$ is $-E\left(\dfrac{\partial^2 \log L}{\partial \theta_i^2}\right)$ and that the covariance of $\dfrac{\partial \log L}{\partial \theta_i}$ and $\dfrac{\partial \log L}{\partial \theta_j}$ is $-E\left(\dfrac{\partial^2 \log L}{\partial \theta_i \partial \theta_j}\right)$. The $k \times k$ matrix containing these variances and covariances, that is to say the matrix in which the element in the ith row and jth column is the covariance of $\dfrac{\partial \log L}{\partial \theta_i}$ and $\dfrac{\partial \log L}{\partial \theta_j}$, is usually called the information matrix and denoted by I. If we consider the k equations of the kind

$$\frac{\partial \log L}{\partial \theta_i} \text{ (evaluated at } \hat{\theta}_1, \hat{\theta}_2, \ldots, \hat{\theta}_k) = 0$$

$$\doteqdot \frac{\partial \log L}{\partial \theta_i} + (\hat{\theta}_1 - \theta_1) \frac{\partial^2 \log L}{\partial \theta_1 \partial \theta_i} + \ldots + (\hat{\theta}_k - \theta_k) \frac{\partial^2 \log L}{\partial \theta_k \partial \theta_i}$$

in which the Expected values of the second derivatives can be substituted for their observed values, it will be seen that $\hat{\theta}_1, \hat{\theta}_2, \ldots, \hat{\theta}_k$ are asymptotically normal, asymptotically unbiased and that their variance-covariance matrix is approximately I^{-1}, the inverse of I.

For example, if n observations have been taken from a normal distribution, then

$$\frac{\partial^2 \log L}{\partial \mu^2} = -n/\sigma^2, \text{ whose Expected value is } -n/\sigma^2$$

$$\frac{\partial^2 \log L}{\partial \mu \partial \sigma^2} = -\sum \frac{(x_i - \mu)}{\sigma^4}, \text{ whose Expected value is } 0$$

$$\frac{\partial^2 \log L}{(\partial \sigma^2)^2} = \frac{n}{2\sigma^4} - \sum \frac{(x_i - \mu)^2}{\sigma^6}, \text{ whose Expected value is } -\frac{n}{2\sigma^4}.$$

Thus the information matrix is

$$I = \begin{bmatrix} \dfrac{n}{\sigma^2} & 0 \\ 0 & \dfrac{n}{2\sigma^4} \end{bmatrix}$$

and the approximate variance-covariance matrix of $\hat{\mu}$ and $\hat{\sigma}^2$ is

$$I^{-1} = \begin{bmatrix} \dfrac{\sigma^2}{n} & 0 \\ 0 & \dfrac{2\sigma^4}{n} \end{bmatrix}.$$

The variance of $\hat{\mu}$ and the covariance are exact; the variance of $\hat{\sigma}^2$ is only approximately correct, which one might expect since $\hat{\sigma}^2$ is not unbiased.

Problems

11.1. If $x_{(1)}, x_{(2)}, \ldots, x_{(n)}$ is a sample of n observations re-arranged in rank order from a continuous distribution with density function $f(x)$ and cumulative probability function $F(x)$, show that the probability that $x_{(r)}$ lies between y and $y+dy$ is

$$\frac{n!}{(r-1)!(n-r)!} F^{r-1}(y)[1-F(y)]^{n-r} f(y)dy$$

[Find the probability of obtaining $r-1$ observations less than y, $n-r$ observations greater than y and one observation between y and $y+dy$ *in a particular order*, and then multiply this probability by the number of ways in which the sample can be re-arranged to give the same *ordered* sample.]

Suppose that the underlying random variable is uniformly distributed between 0 and 1 and that n is odd. Show that the sample median follows the Beta distribution (see Problem 8.8), evaluate its mean and variance, and compare the latter with the approximate value derived from the formula on p. 190.

11.2. Show that the mean and median of a random sample of $2n+1$ observations from the uniform distribution

$$f(x) = \frac{1}{2\mu} \qquad 0 \leqslant x \leqslant 2\mu$$

are both unbiased estimators of the population mean μ.

What are their standard errors? The largest element of the sample is y. Show that $(n+1)y/(2n+1)$ is also an unbiased estimator of μ and find its standard error. Comment. [Certificate, 1964. Note that Fisher's theorem on minimum variance estimators cannot be applied since the range depends on μ.]

11.3. Suppose that a suspension of bacteria contains μ bacteria per c.c. and that for a sample drawn from this suspension it is only possible to determine whether it is sterile because it contains no bacteria or non-sterile because it contains at least one bacterium. If n samples each of x c.c. are tested and a of them are found to be sterile and $n-a$ non-sterile find the maximum likelihood estimator of μ and its standard error. What is the best value of x?

11.4. Suppose that we wish to minimise $\phi(x_1, x_2, \ldots, x_n)$ subject to the k constraints $f_1(x_1, \ldots, x_n) = \ldots = f_k(x_1, \ldots, x_n) = 0$. This is most easily done by the method of undetermined Lagrangian multipliers, that is to say by solving the $n+k$ equations

$$\frac{\partial \phi}{\partial x_i}+\frac{\lambda_1 \partial f_1}{\partial x_i}+\ldots+\frac{\lambda_k \partial f_k}{\partial x_i}=0 \qquad i=1, \ldots, n$$

$$f_1=f_2=\ldots=f_k=0$$

in the $n+k$ unknowns $x_1, x_2, \ldots, x_n, \lambda_1, \lambda_2, \ldots, \lambda_k$ (see Courant, 1934).

Suppose that m unbiased independent estimates, $t_1, t_2, \ldots, t_m$, have been made of some parameter θ, having variances $\sigma_1^2, \sigma_2^2, \ldots, \sigma_m^2$. Find the unbiased linear combination of these estimates, $\Sigma w_i t_i$, which has minimum variance. What is this minimum variance? What happens if t_i is a mean based on n_i observations from a population with variance σ^2?

11.5. One way of estimating the size of an animal population is to capture, mark and release M animals out of the total population of unknown size N and later, when the marked animals have mixed freely with the unmarked, to catch a further sample of size n of which m are found to be marked. If all animals are equally likely to be caught m will follow the hypergeometric distribution (Problems 6.7 and 6.8) since sampling is without replacement, but if M is small compared with N, as is usually the case, this can be approximated by the more tractable binomial distribution. Show that, under the latter simplification, the maximum likelihood estimator of N is the obvious estimate Mn/m. However, both the mean and the variance of this estimator are infinite since m has a non-zero probability of being zero! It is therefore preferable to estimate $1/N$ by the better-behaved estimator m/nM, which is an unbiased estimator of $1/N$ and whose variance can be exactly calculated. Calculate this variance and show how approximate confidence limits can be constructed for $1/N$ and hence for N. Calculate a 95 per cent confidence interval for N in this way when $M=n=100$, $m=10$.

11.6. In the above problem it was assumed that the size of the second sample was fixed and that in repeated samples of this size the number of marked animals would therefore follow a hypergeometric distribution. An alternative procedure is to continue sampling until a fixed number of marked animals has been caught, so that the total sample size n now becomes the random variable; this method is known as inverse sampling. Write down the probability function of n and evaluate its mean and variance by finding $E(n)$ and $E[n(n+1)]$. Hence show that nM/m is a biased estimator of N, find an unbiased estimator of N of the same general type and evaluate its sampling variance. [See Bailey, 1951]

11.7. Suppose that $n_1, n_2, \ldots, n_k$ are the numbers of observations falling into k exclusive classes and that P_i, the probability of an observation falling in the ith class, is a function of a single unknown parameter θ.

Show that

$$\frac{d \log L}{d\theta} = \sum_{i=1}^{k} \frac{n_i}{\pi_i} \frac{d\pi_i}{d\theta}$$

$$I(\theta) = -E\left(\frac{d^2 \log L}{d\theta^2}\right) = n \sum_{i=1}^{k} \frac{1}{\pi_i} \left(\frac{d\pi_i}{d\theta}\right)^2$$

Given the model of Problem 2.5 and the data of Table 7 on p. 25 find the maximum likelihood estimator of θ, find confidence limits for θ and hence find confidence limits for π, the probability of crossing over, where $\theta = (1 - \pi)^2$.

11.8. It sometimes happens that the maximum likelihood equation does not have an explicit solution. In this case it can be solved by an iterative procedure usually known as the method of scoring. Let us write $S(\theta_0)$ for the value of $\frac{d \log L}{d\theta}$ evaluated at $\theta = \theta_0$; this quantity is called the maximum likelihood score at θ_0. To solve the equation $S(\theta) = 0$, we evaluate the score and its first derivative at some trial value, θ_0, thought to be near $\hat{\theta}$ and then calculate a corrected value, θ_1, from the approximate formula

$$S(\theta_1) \doteq S(\theta_0) + (\theta_1 - \theta_0) S'(\theta_0)$$

If we put $S(\theta_1) = 0$, we find that

$$\theta_1 = \theta_0 - \frac{S(\theta_0)}{S'(\theta_0)}$$

We now evaluate the score at θ_1 and repeat the above process until it converges. It may be simpler to use the Expected rather than the observed value of the derivation of the score, in which case the above formula becomes

$$\theta_1 = \theta_0 + \frac{S(\theta_0)}{I(\theta_0)}$$

where

$$I(\theta_0) = -E\left(\frac{d^2 \log L}{d\theta^2}\right)_{\theta = \theta_0} = -E[S'(\theta_0)].$$

$I(\theta_0)$ is called the amount of information at θ_0.

Assume that you were unable to find an explicit solution for the maximum likelihood equation in the previous problem and solve it by the above iterative method starting with the trial value $\theta_0 = \frac{1}{4}$.

Chapter 12

REGRESSION AND CORRELATION

The scientist is often interested in the relationship between two, or more, variables; for simplicity we shall suppose that there are only two. There are, broadly speaking, two main types of problem. In the first, which usually gives rise to a *regression analysis*, interest is focused on one of the variables which is thought of as being *dependent* on the other, the *independent* variable. For example, the velocity of a chemical reaction usually varies with the temperature at which it is taking place; in many cases the relationship follows Arrhenius' equation

$$\text{log velocity} = A - \frac{B}{T}$$

where T is the absolute temperature and A and B are constants typical of the reaction; a graph of log velocity against the reciprocal of the absolute temperature will therefore give a straight line with a negative slope. Reaction velocity is here the dependent and temperature the independent variable since it is the rise in temperature which causes the increase in velocity and not the other way round.

In the second type of problem, which usually leads to the calculation of a *correlation coefficient*, neither of the variables can be singled out as of prior importance to the other and one is interested in their interdependence rather than in the dependence of one of them on the other. For example, if one had measured the length and breadth of the head in a group of men, one might want to know to what extent the two were associated although neither of them is causally dependent on the other.

The theory of regression has many points in common with that of correlation, although they answer rather different questions. It is now recognised that regression techniques are more flexible and can answer a wider range of questions than correlation techniques which are used less frequently than they once were. We shall consider regression first.

LINEAR REGRESSION AND THE METHOD OF LEAST SQUARES

As a typical regression problem consider the data in Table 22 on the comb-growth (increase in length+height of the comb) in 5 groups of 5 capons (castrated cocks) receiving different doses of androsterone (male sex hormone) (Greenwood *et al.*, 1935). It will be seen from Fig. 30, in which comb-growth is plotted against the logarithm of the dose, that there is an approximately linear relationship between these two quantities over the range of doses used. Comb-growth is obviously the dependent, and dose of androsterone the independent, variable.

TABLE 22

Comb-growth in capons receiving different doses of androsterone

Dose (mg. androsterone)	$\frac{1}{2}$	1	2	4	8
Log_2 dose (x)	−1	0	1	2	3
	8	5	13	17	17
	1	6	7	14	17
Comb-growth (mm.) (y)	1	9	12	14	20
	3	7	10	19	18
	1	4	11	13	15

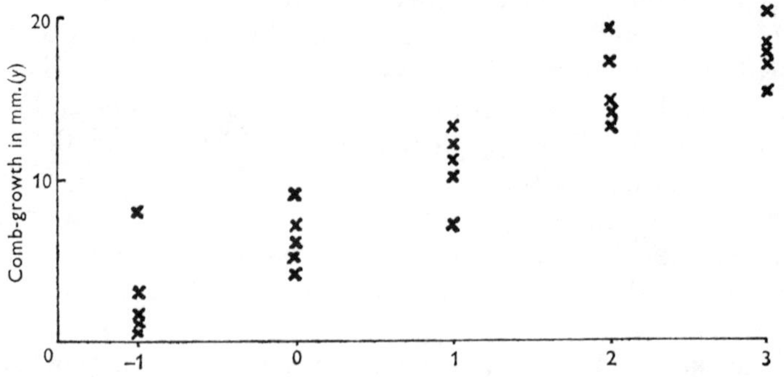

FIG. 30. Comb-growth in capons receiving different doses of androsterone

It is clear that, for a fixed value of the dose, the comb-growth varies considerably from one bird to another and may be regarded as a random variable with a mean, a variance and so on, which will be symbolised by $E(Y \mid x)$, $V(Y \mid x)$, etc., where Y stands for the comb-growth (the dependent variable) and x for $\log_2$ dose (the independent variable); it should be noticed that whereas Y is a random variable once x has been fixed, x is not a random variable but is fixed by and known exactly to the experimenter. The characteristics of the distribution of Y for a given value of x, and in particular $E(Y \mid x)$, are functions of x and may be expected to change with x. The graph of $E(Y \mid x)$ as a function of x is called the *regression* of Y on x; the purpose of regression analysis is to make inferences about the form of this graph.

The simplest and most important type of regression is the straight line

$$E(Y \mid x) = \alpha + \beta x$$

where β is the slope of the line and α its intercept at $x = 0$. As we remarked above, the regression of comb-growth on log dose seems to be approximately linear within the range of doses from $\frac{1}{2}$ mg. to 8 mg.; it cannot, however, be linear over the entire range of doses since (1) there must be an upper limit to the comb-growth as the dose is indefinitely increased, and (2) the comb-growth for zero dose, when log dose $= -\infty$, must be zero and not $-\infty$! This example illustrates the danger of extrapolation.

Let us suppose then that we have n pairs of observations, (x_1, y_1), (x_2, y_2), ..., (x_n, y_n), on two variables of which x is the independent and y the dependent variable and that we wish to estimate the regression of y on x which is assumed to be linear,

$$E(Y \mid x) = \alpha + \beta x.$$

The standard procedure is to choose as the estimated regression line

$$y = a + bx$$

that line which minimises the sum of the squared deviations

of observed from estimated values of y, that is to say the line which minimises the quantity

$$S^2 = \sum_{i=1}^{n} (y_i - a - bx_i)^2.$$

These deviations are shown graphically in Fig. 31. This method is known as the *method of least squares.* It was first considered in connection with errors of astronomical observations by Legendre in 1806 and by Gauss in 1809.

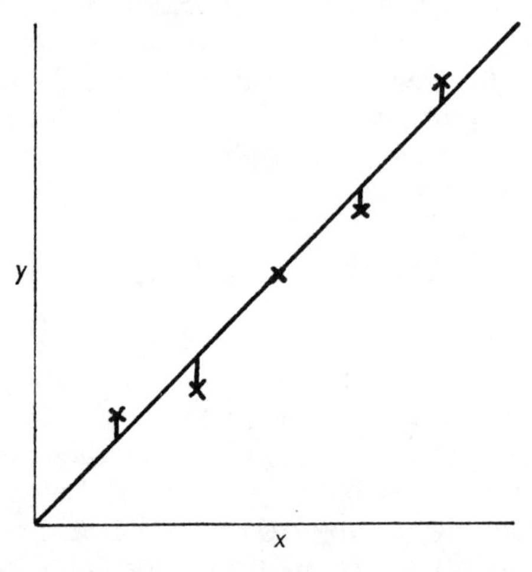

FIG. 31. The 'best' line is the line which minimises the sum of the squares of the deviations in the direction shown

To find the line which minimises S^2 we must solve the pair of simultaneous equations:

$$\frac{\partial S^2}{\partial a} = -2\sum(y_i - a - bx_i) = -2\sum y_i + 2na + 2b\sum x_i = 0$$

$$\frac{\partial S^2}{\partial b} = -2\sum x_i(y_i - a - bx_i) = -2\sum x_i y_i + 2a\sum x_i + 2b\sum x_i^2 = 0.$$

The solution of the first equation is

$$a = \bar{y} - b\bar{x}$$

which tells us that the line passes through the point $(\bar{x}, \bar{y})$. Substituting this expression in the second equation we find

$$b = \frac{\sum x_i y_i - n\bar{x}\bar{y}}{\sum x_i^2 - n\bar{x}^2} = \frac{\sum (x_i - \bar{x})(y_i - \bar{y})}{\sum (x_i - \bar{x})^2}.$$

It is also convenient to have a formula for S^2, the residual sum of squares which has been minimised. We have

$$\begin{aligned} S^2 &= \sum (y_i - a - bx_i)^2 = \sum [(y_i - \bar{y}) - b(x_i - \bar{x})]^2 \\ &= \sum (y_i - \bar{y})^2 + b^2 \sum (x_i - \bar{x})^2 - 2b \sum (y_i - \bar{y})(x_i - \bar{x}) \\ &= \sum (y_i - \bar{y})^2 - b^2 \sum (x_i - \bar{x})^2. \end{aligned}$$

The second term in the last expression represents the contribution to the variability of the y's which has been removed by calculating the regression.

We must now consider the justification of this method of estimation. We suppose that the dependent variable is normally distributed with a variance σ^2 which does not depend on x. Our model is then

$$y_i = \alpha + \beta x_i + \epsilon_i$$

where ϵ_i, the random error in the ith observation, is normally distributed with zero mean and variance σ^2. The logarithm of the likelihood of the observations is

$$\log L = -\tfrac{1}{2} n \log 2\pi - \tfrac{1}{2} n \log \sigma^2 - \tfrac{1}{2} \sum (y_i - \alpha - \beta x_i)^2 / \sigma^2.$$

Since α and β occur only in the third term, the maximum likelihood estimates of these parameters are found by minimising that term and are thus the same as the least squares estimates. The maximum likelihood estimator of σ^2 is S^2/n. It is also quite easy to show from their sampling distributions (see Appendix) that these three estimators are jointly sufficient for the three parameters of the distribution.

The sampling distributions of a, b and S^2 are investigated in the Appendix to this chapter. It must be remembered that only the dependent variable is considered as a random variable so that these distributions are obtained by imagining that repeated samples of size n are taken with the same, constant values of the x_i's but with different values of the ϵ_i's and hence of the y_i's. It is shown that a and b are normally distributed, unbiased estimators of α and β respectively with variances given by the formulae:

$$V(a) = \sigma^2\left\{\frac{1}{n} + \frac{\bar{x}^2}{\Sigma(x_i-\bar{x})^2}\right\}$$

$$V(b) = \frac{\sigma^2}{\sum(x_i-\bar{x})^2}.$$

These variances are the minimum possible variances of any unbiased estimators of α and β. Furthermore, S^2/σ^2 is distributed as a χ^2 variate with $n-2$ degrees of freedom independently of a and b, so that $s^2 = S^2/(n-2)$ is an unbiased estimator of σ^2. The number of degrees of freedom is 2 less than the number of observations because 2 parameters other than σ^2 have been estimated from the data. The distributions of $\bar{y}$ and b are independent but the distributions of a and b are not independent unless $\bar{x} = 0$, when $a = \bar{y}$. When $\bar{x}$ is positive a and b are negatively correlated, which means that an over-estimate of β is more likely than not to be accompanied by an under-estimate of α, and *vice versa*; when $\bar{x}$ is negative the contrary is true.

These results can be used to perform significance tests or to construct confidence intervals by means of the t distribution. For example,

$$\frac{(b-\beta)}{s}\sqrt{\Sigma(x_i-\bar{x})^2}$$

follows the t distribution with $n-2$ degrees of freedom. This fact can be used either to test a particular value of β, such as $\beta = 0$, which means that there is no relationship between the variables, or to place a confidence interval on β. Inferences about α can be made in a similar way.

It has been assumed that the random errors in the dependent variable are normally distributed with the same variance. This assumption may be wrong in two ways. First, the underlying distribution may not be normal. In this case a and b are no longer normally distributed, but their Expected values and variances are unchanged; S^2/σ^2 no longer follows the χ^2 distribution but its Expected value is still $n-2$. Second, the variance of Y may not be constant but may depend on x; the regression is then said to be *heteroscedastic* (from the Greek meaning 'different scatter'). In this case a and b are still unbiased estimators and are normally distributed if the underlying distribution is normal, but the formulae for their variances require modification. If the form of the relationship between the variance of Y and x is known, for example if the variance is known to be proportional to x, more efficient estimators can be obtained by weighting the observations with weights inversely proportional to their variances. In general, however, small departures from normality or homoscedasticity will have little effect on inferences about the regression line and may be ignored.

Curvilinear and Multiple Regression

It has been assumed so far that the regression is linear; it is clearly important to be able to test the adequacy of this hypothesis. There are two ways in which this may be done.

Consider the data on the comb-growth of capons in Table 22. We shall change our notation slightly and write y_{ij} for the response of the jth bird at the ith dose level (e.g. $y_{35} = 11$) and $y_{i.}$ for the average response to the ith level (e.g. $y_{2.} = 6{\cdot}2$); note that i and j both run from 1 to 5. If we write the deviation of an observed value, y_{ij}, from its estimated value, $a+bx_i$, in the form

$$y_{ij}-a-bx_i = (y_{i.}-a-bx_i)+(y_{ij}-y_{i.})$$

the residual sum of squares can be split up into two components:

$$\begin{aligned} S^2 = \sum_i\sum_j (y_{ij}-a-bx_i)^2 &= 5\sum_i (y_{i.}-a-bx_i)^2+\sum_i\sum_j (y_{ij}-y_{i.})^2 \\ &= \qquad S_1^2 \qquad\qquad + \qquad S_2^2. \end{aligned}$$

The factor 5 occurs in S_1^2 because of summation over the index j; the cross-product term vanishes because $\sum_j(y_{ij}-y_{i.})$ is zero for all i.

S_2^2 is the sum of the squared deviations of the observations from their respective means; hence S_2^2/σ^2 is a χ^2 variate with 20 degrees of freedom, regardless of the shape of the regression, since it is the sum of 5 sums of squares each with 4 degrees of freedom. It follows that *if* the regression is linear, S_1^2/σ^2 is a χ^2 variate with 3 degrees of freedom since in that case S^2/σ^2 is a χ^2 variate with 23 degrees of freedom. (This can also be proved directly from the fact that $5\sum_i(y_{i.}-\alpha-\beta x_i)^2/\sigma^2$ is a χ^2 variate with 5 degrees of freedom; note that $y_{i.}$ is normally distributed with mean $\alpha+\beta x_i$ and variance $\sigma^2/5$.) Hence, if the regression is linear, the mean squares, $M_1 = S_1^2/3$ and $M_2 = S_2^2/20$, are independent and unbiased estimators of σ^2; if the regression is not linear the distribution of M_2 will be unchanged but M_1 will have a larger Expected value. A test for linearity can therefore be obtained by calculating the ratio $F = M_1/M_2$. If the regression is linear this quantity will follow the F distribution with 3 and 20 degrees of freedom, and we should expect it to be approximately 1; if the regression is not linear its Expected value will be increased.

For the data in Table 22, $S_1^2 = 9{\cdot}76$ and $S_2^2 = 111{\cdot}20$, whence $M_1 = 3{\cdot}25$, $M_2 = 5{\cdot}56$ and $F = 3{\cdot}25/5{\cdot}56 = {\cdot}58$. Since F is less than 1 there is clearly no reason to reject the null hypothesis that the regression is linear; if F had been larger its significance could have been evaluated from tables of the percentage points of the F distribution.

This method of testing for linearity, which is typical of the sort of argument used in the Analysis of Variance, can only be employed when several observations on the dependent variable are made for each value of the independent variable. Very often, however, each x value occurs only once and another method must be found. In this case the following argument can be employed. It is unlikely that the regression is highly non-linear; for if it were it would be obvious on inspection and we should not be using a linear model. It is therefore reasonable to suppose that the true regression can be approximated to a good degree of accuracy by a quadratic curve:

$$E(Y \mid x) = \alpha+\beta x+\gamma x^2.$$

A test for linearity can therefore be constructed by finding the best fitting quadratic curve by the method of least squares and then testing the hypothesis that $\gamma = 0$ by a t test. The variance of c, the least squares estimator of γ, can be calculated by an extension of the method used in the Appendix for calculating the variances of α and β in a linear model; the residual sum of squares from the quadratic curve has $n-3$ degrees of freedom. For further details the reader is referred to a book on regression analysis such as Williams (1959).

The method of least squares can be used in a similar way to calculate a cubic or a quartic regression, or in general a polynomial of any degree. It can also be extended to situations in which there are two or more independent variables. Suppose that we wish to investigate the relationship between the yield of some crop and the rainfall and average temperature in different years; we denote the values of these quantities in the ith year by y_i, x_{1i} and x_{2i} respectively. There are now two independent variables and a simple linear model would be

$$y_i = \alpha+\beta_1 x_{1i}+\beta_2 x_{2i}+\epsilon_i$$

where, as before, ϵ_i is a random error. The three parameters, α, β_1 and β_2, can be estimated by the method of least squares and significance tests or confidence intervals can be constructed. (For details of the calculations involved see Williams, 1959, or Bailey, 1959.) If $\beta_1 = 0$ rainfall has no effect and if $\beta_2 = 0$ temperature has no effect; if these parameters are not zero, their values will provide information about the importance of these two factors.

It should be noted that a regression of yield on rainfall alone will not necessarily give the correct information about the importance of this factor. Suppose that temperature is the important factor and that rainfall has little effect within the range of variability represented. It is likely that hot years will be dry years and that cold years will be wet. If the yield is higher in hot years than in cold years it will be found, when we plot yield against rainfall, that the yield is apparently higher in dry years than in wet years. This relationship may be entirely spurious, due to the fact that dry years are hot years and *vice versa*. The only way to find out which of these two factors is the real causal factor is to calculate the joint regression of yield on rainfall and temperature simultaneously. Even then it is always possible that the real causal factor is some third, unconsidered variable correlated with either rainfall or temperature. The possibility of spurious association should always be borne in mind in interpreting non-experimental data; this danger can be avoided in experimental work by proper randomisation.

The Effect of Errors in the Independent Variable

It has been assumed so far that the independent variable is known exactly to the experimenter and is not subject to random error. This assumption is frequently not satisfied and it is therefore important to investigate the effect of errors in the independent variable on the regression line. To fix our ideas we shall consider the regression of the height of sons on that of their parents.

Height can be regarded as composed of a genetic contribution on which is superimposed a random element due to environmental differences. Let us first suppose that the environmental component is negligible so that actual height is the same as genetic height. If both parents make equal genetic contributions, then the son's height will on the average be equal to the mean height of his parents; that is to say, the regression of son's height (S) on the heights of his father (F) and mother (M) will be

$$S = \tfrac{1}{2}\mathrm{F} + \tfrac{1}{2}M.$$

There will of course be some random variability about this regression line because each parent contributes only half his genes to his offspring. Furthermore, if parents marry at random with respect to height, that is to say if there is no tendency for tall men to marry tall women, the regression of son's height on father's height will be

$$S = \tfrac{1}{2}\mu + \tfrac{1}{2}F$$

where μ is the average height in the population. Likewise the regression of son's height on mother's height will be

$$S = \tfrac{1}{2}\mu + \tfrac{1}{2}M.$$

(For the sake of simplicity, sex differences in height have

been ignored.) If, however, there is a tendency for tall men to marry tall women, then a knowledge of the height of one parent enables us to predict something about the height of the other parent, and so the regression coefficient will be increased; that is to say the regression of son's height on father's height will be

$$S = \alpha + \beta F$$

where β is greater than $\frac{1}{2}$; α must then be less than $\frac{1}{2}\mu$ to maintain the average height of all sons constant. In the extreme case when there is a perfect correlation between the heights of husband and wife, that is to say when men marry women of exactly their own height, the regression of son's height on father's height will be

$$S = F.$$

Let us now consider the effect of the environmental component on these regressions; actual height is now to be regarded as the sum of genetic height and a random error contributed by the environment; we may define the environmental contribution in such a way that its mean value is zero. The effect of this environmental contribution to the son's height will clearly be to increase the scatter about the regression lines but to leave the lines themselves unaltered. The environmental contribution to the heights of the parents, on the other hand, will not only increase the scatter about the line but will decrease its slope. A tall father, for example, may be tall for one of two reasons; his genetic height may be large, or the environmental contribution may be large. In fact both these factors are likely to be large, but it can be seen that on the average tall fathers are more likely to have large, positive environmental contributions than short fathers. It follows that the average genetic height of a group of tall fathers must be less than their actual height since their average environmental contribution is greater than zero. Similarly the average genetic height of a group of short fathers will be greater than their actual height. Hence the slope of the regression of son's height on father's actual height will be less than that of the regression on his genetic height.

A classical paper on the inheritance of height in man was

published by Pearson and Lee in 1903. They found that the joint regression of son's height, in inches, on the heights of both parents was

$$S = 14 + \cdot 41F + \cdot 43M.$$

This result was based on over a thousand observations, so that there can be no doubt about its reliability. The regression coefficients are significantly below the value of $\frac{1}{2}$ predicted on the assumption that height is entirely under genetic control; it must be concluded that there is an appreciable environmental effect. The single regressions on father's height and mother's height were respectively

$$S = 34 + \cdot 52F$$

and

$$S = 34 + \cdot 56M.$$

The regression coefficients are higher than the corresponding coefficients in the double regression because the heights of the two parents are correlated; the correlation coefficient was in fact ·25.

Suppose, to take another example, that a new method has been developed to measure the concentration of some substance in solution and that in order to test its validity measurements, $y_1, y_2, \ldots, y_n$, have been made by this method on n solutions of known concentrations, $x_1, x_2, \ldots, x_n$. If the method provides an unbiased measure of the concentration, the regression of y on x should be a straight line with unit slope passing through the origin. If, however, the x_i's are not known concentrations but are themselves estimates by another method of known validity but subject to random error, the regression of y on x will have a slope less than unity and will cut the y axis above the origin; the constants of this line cannot therefore be used to test the validity of the new method.

The presence of error in the independent variate thus decreases the slope of the regression line. Whether or not this matters depends on why one is interested in the regression. If one wants to know the underlying relationship between the dependent and the independent variables, that is to say the relationship which would exist between them if there were no random error in either of them, then the presence of error in the independent variable destroys our information about this relationship. If we knew the relative magnitudes of the variability of the two variables we could calculate by how much the slope had been decreased by the errors in the independent variate and hence make a correction to our estimate of the slope. Usually, however, we do not know the relative magnitudes of the two sources of error. In these circumstances the underlying relationship is unidentifiable however many observations are made, and the best that can be done is to argue that the underlying relation must lie somewhere between the regression line of y on x and that of x on y; the former would be the appropriate regression to consider if all the variability could be attributed to random errors in y and the latter if it were all due to errors in x. These two regressions are of course not the same; the relationship between them will be considered in the next section.

The presence of random errors in both variates makes the underlying relationship between them unidentifiable. If, however, one wants to use the regression line only for purposes of prediction this does not matter. For example, the regression of son's height on the height of his parents

$$S = 14 + \cdot 41F + \cdot 43M$$

is not a valid estimate of the regression of son's height on the genetic height of his parents. It is nevertheless a valid formula for predicting the height of a son given the heights of his father and mother. If one is interested in regressions for prediction purposes rather than for testing some theoretical model about the underlying relationship, the presence of error in the independent variate does not matter.

Correlation

We shall now consider the case when both variables are random variables having some joint probability distribution; as a typical example we shall take the joint distribution of head breadth and head length in Table 10 on p. 40. The regression of Y on X is defined as the conditional Expected value of Y given that $X = x$, or in symbols $E(Y|X = x)$; this is the limiting value of the curve obtained by plotting the average value of y for different values of x. For example, the crosses in Fig. 32 represent the average head breadth for fixed head length; it is clear that the regression of head breadth on head length is very nearly linear, and that the linear regression calculated by the method of least squares passes almost exactly through the crosses. The other line in Fig. 32 is the linear

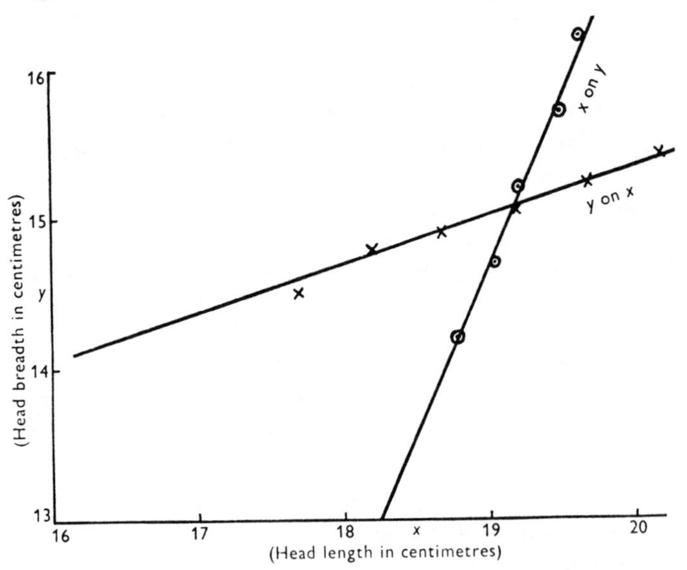

FIG. 32. The regressions of y on x and of x on y for the data of Table 10

regression of head length on head breadth, and the circles represent the average head length for fixed head breadth, the curve through which tends to the regression of X on Y. This figure illustrates the very important point that the regression of Y on X and of X on Y are not the same; we should use the former regression if we wanted to use x to predict y and the latter if we wanted to use y to predict x.

If $y = a+bx$ is the linear regression of y on x calculated by the method of least squares and if $x = a'+b'y$ is the corresponding regression of x on y, then

$$bb' = \frac{[\Sigma(x_i-\bar{x})(y_i-\bar{y})]^2}{\Sigma(x_i-\bar{x})^2\Sigma(y_i-\bar{y})^2} = r^2.$$

(See p. 74 for the definition of the correlation coefficient, r.) Thus if $r^2 = 1$ the points lie exactly on a straight line and so the two regression lines are the same. Otherwise, however, bb' is less than 1 and so the two lines do not coincide; for the condition that they coincide is that $b' = 1/b$ or $bb' = 1$. The smaller r^2, the larger is the angle between the two lines until, when $r^2 = 0$, they are at right angles since the regression of y on x is parallel to the x-axis and that of x on y is parallel to the y-axis.

The concept of regression can therefore be meaningfully applied to a bivariate distribution, but it is often most useful to calculate some measure of the degree of relationship between the variables. The correlation coefficient is used most frequently and we must now consider what this coefficient measures. If we fit a linear regression of y on x, $y = a+bx$, the residual sum of squares of the y's about this line is

$$S^2 = \sum(y_i-\bar{y})^2 - b^2\sum(x_i-\bar{x})^2.$$

The second term on the right-hand side is the sum of squares removed by fitting the line. It can easily be shown by substituting the appropriate formula for b that

$$r^2 = \frac{b^2\sum(x_i-\bar{x})^2}{\sum(y_i-\bar{y})^2}.$$

The square of the correlation coefficient is thus the sum of squares removed by estimating b expressed as a proportion of the original sum of squares. It follows from a similar argument that r^2 can also be considered as the proportion of the sum of squares of the x's removed by fitting the regression of x on y, $x = a'+b'y$. The correlation coefficient does not give a privileged position to either of these variables. It is clear that r^2 must lie between 0 and 1 so that r must lie between -1 and $+1$. The correlation coefficient is negative when b is negative and positive when b is positive. It takes the extreme values $+1$ or -1 when the points lie exactly on a straight line and it is zero when b is zero. It is thus a reasonable measure of the degree of association between the two variables provided that there is a linear relationship between them.

We must now consider how the observed value, r, can be used to perform significance tests and find confidence intervals for the true value, ρ. (See p. 74 for the definition of ρ.) We shall first construct a significance test to test whether or not $\rho = 0$. We shall suppose that X and Y are in fact independently and normally distributed; if this is so then $\rho = 0$, although the converse is not necessarily true. Under these circumstances the quantity

$$\frac{r\sqrt{n-2}}{\sqrt{1-r^2}}$$

should follow the t distribution with $n-2$ degrees of freedom. To test the hypothesis that $\rho = 0$ we therefore calculate this quantity and then find the probability of obtaining a value as large as or larger than the observed value from tables of the t distribution. A two-tailed test will usually be appropriate but we could use a one-tailed test if we knew, for example, that ρ could not be negative.

(To prove that this quantity should follow the t distribution we observe that, if we calculate the regression of y on x, then

$$\frac{\sum(x_i-\bar{x})^2 b^2}{\sum(y_i-\bar{y})^2} = \frac{\sum(x_i-\bar{x})^2[\sum(x_i-\bar{x})(y_i-\bar{y})]^2}{\sum(y_i-\bar{y})^2[\sum(x_i-\bar{x})^2]^2} = r^2$$

and

$$\frac{S^2}{\sum(y_i-\bar{y})^2} = 1 - \frac{\sum(y_i-\bar{y})^2 - S^2}{\sum(y_i-\bar{y})^2} = 1-r^2.$$

If we divide the first expression by the second, multiply by $n-2$ and then take the square root we find that

$$\frac{r\sqrt{n-2}}{\sqrt{1-r^2}} = \frac{b\sqrt{\sum(x_i-\bar{x})^2}}{s}.$$

We have already seen that, for fixed values of the x_i's, the quantity on the right-hand side follows the t distribution with $n-2$ degrees of freedom when $\beta = 0$. Since it follows this distribution for *any* fixed values of the x_i's it must follow the same distribution when they become random variables.)

We must now consider the sampling distribution of r when ρ

is not zero. We shall suppose that the marginal distributions of X and Y are normal and that the regression of Y on X is linear, that is to say that the conditional Expected value, $E(Y \mid X = x)$, is a linear function of x. (If the regression is not linear the correlation coefficient is not a useful measure of the degree of association between the two variables.) Under these circumstances the joint distribution of X and Y is the bivariate normal distribution (see Problem 12.4.) An explicit expression can be found for the sampling distribution of a correlation coefficient calculated from a sample of size n from this distribution but it is rather complicated and its shape depends on ρ, being skew to the left when ρ is positive and skew to the right when ρ is negative. Fisher (1921) has shown that the transformed variate $z = \frac{1}{2} \log \frac{1+r}{1-r}$ is approximately normally distributed with mean $\zeta = \frac{1}{2} \log \frac{1+\rho}{1-\rho}$ and variance $1/(n-3)$. This distribution is much easier to handle than the distribution of r since only its mean depends on ρ and since tables of the normal distribution are readily available; Fisher's z transformation is tabulated in most statistical tables.

The possibility of spurious correlation should always be borne in mind in interpreting correlation coefficients. Thus it has been shown that there is a high correlation between mortality from heart disease and the annual number of television licences issued. It would be wrong to conclude that watching television causes heart disease; the correlation is due to the fact that both heart disease and television viewing have increased over the years, though for different reasons. On the other hand, the correlation between lung cancer and smoking almost certainly reflects a causal connection between the two; a great deal of special pleading is required to explain this correlation as due to some third factor, such as a hereditary predisposition, which causes both lung cancer and a tendency to smoke excessively. One can never be certain in non-experimental situations that a correlation reflects a true causal connection but common sense must be used in deciding whether or not this is likely.

Appendix

Sampling distributions of linear regression estimates

Our model is that

$$y_i = \alpha + \beta x_i + \epsilon_i \qquad i = 1, 2, \ldots, n$$

where the ϵ_i's are normally and independently distributed with zero mean and variance σ^2. We consider first the sampling distribution of b. The numerator of b is

$$\sum(x_i - \bar{x})(y_i - \bar{y}) = \sum(x_i - \bar{x})\, y_i = \sum(x_i - \bar{x})(\alpha + \beta x_i + \epsilon_i)$$
$$= \beta\sum(x_i - \bar{x})^2 + \sum\epsilon_i(x_i - \bar{x}).$$

(Note that $\sum(x_i - \bar{x})$, and hence any constant multiple of this quantity, is zero.) Hence

$$b = \frac{\sum(x_i - \bar{x})(y_i - \bar{y})}{\sum(x_i - \bar{x})^2} = \beta + \frac{\sum\epsilon_i(x_i - \bar{x})}{\sum(x_i - \bar{x})^2}$$

b is thus normally distributed since it is the sum of n independent normal variates. Furthermore $E(b) = \beta$ since $E(\epsilon_i) = 0$ for all i, and the variance of b is the Expected value of the square of the second term, which is $\sigma^2/\sum(x_i - \bar{x})^2$. (Only the coefficients of terms like ϵ_i^2 need be considered since $E(\epsilon_i \epsilon_j) = 0$ for $i \neq j$.) Thus b is normally distributed and is an unbiased estimator of β with the above variance.

We now turn to the distribution of a. We first note that

$$\bar{y} = \sum y_i/n = \alpha + \beta\bar{x} + \sum\epsilon_i/n.$$

It follows that $\bar{y}$ is normally distributed with mean $\alpha + \beta\bar{x}$ and variance σ^2/n. Furthermore the covariance of $\bar{y}$ and b is

$$E\left[\frac{\sum\epsilon_i}{n} \cdot \frac{\sum\epsilon_i(x_i - \bar{x})}{\sum(x_i - \bar{x})^2}\right]$$

which is zero since $\sum(x_i - \bar{x}) = 0$. Hence $a = \bar{y} - b\bar{x}$ is normally distributed with mean

$$E(a) = E(\bar{y}) - \bar{x}E(b) = \alpha$$

and variance

$$V(a) = V(\bar{y})+\bar{x}^2 V(b) = \sigma^2\left[\frac{1}{n}+\frac{\bar{x}^2}{\sum(x_i-\bar{x})^2}\right].$$

Thus a is an unbiased estimator of α with the above variance.

Finally we consider the distribution of the residual sum of squares, S^2. We first note that

$$\epsilon_i = (y_i-\alpha-\beta x_i)$$
$$= (y_i-a-bx_i)+(\bar{y}-\alpha-\beta\bar{x})+(b-\beta)(x_i-\bar{x}).$$

Squaring and summing over i we find that

$$\Sigma\epsilon_i^2 = S^2+n(\bar{y}-\alpha-\beta\bar{x})^2+(b-\beta)^2\sum(x_i-\bar{x})^2.$$

If we divide both sides by σ^2, the left-hand side becomes a χ^2 variate with n degrees of freedom and the second and third terms on the right-hand side are independent χ^2 variates with 1 degree of freedom; they are independent because b and $\bar{y}$ are uncorrelated. It follows from the argument developed in Chapter 8 that S^2/σ^2 is a χ^2 variate with $n-2$ degrees of freedom and is distributed independently of a and b.

Exercises

12.1. From the data in Table 22 on p. 210 calculate Σx_i, Σy_i, Σx_i^2, Σy_i^2 and $\Sigma x_i y_i$ and hence find the equation of the best-fitting straight line. [In evaluating sums like Σx_i it should be remembered that each dose level occurs five times.] Reproduce Fig. 30 and plot this line on it.

12.2. Find the residual sum of squares in the above example (*a*) directly by squaring the deviations of the observed from the predicted points, (*b*) from the formula on p. 213.

12.3. Find 95 per cent confidence intervals for α and β in the above example.

12.4. Find the regressions of head breadth on head length and of head length on head breadth for the data in Table 10 on p. 40 (see Exercises 4.6, 4.7 and 5.5). Find also the mean head breadth for fixed head lengths and *vice versa* and hence reconstruct Fig. 32.

12.5. Find a 95 per cent confidence interval for the correlation between head breadth and head length (see Exercise 5.5).

12.6. Use the data in Table 19 on p. 149 to find the correlation between the hours of sleep gained by the use of hyoscyamine and of hyoscine. Test whether this correlation is significantly different from zero. Would you expect these variables to be correlated?

Problems

12.1. Suppose that n paired observations have been made obeying the regression model

$$y_i = \alpha + \beta x_i + \varepsilon_i \qquad i = 1, 2, \ldots, n$$

where the ε_i's are normally and independently distributed with zero mean and variance σ^2. A further observation, x_{n+1}, is made and is used to predict the corresponding value of y from the formula $y^*_{n+1} = a + bx_{n+1}$. Find the variance of $(y_{n+1} - y^*_{n+1})$ and hence show how a 95 per cent confidence interval for y_{n+1} can be constructed.

12.2. Suppose that $y_i = \alpha + \beta x_i + \varepsilon_i$, where the ε_i's are independently but not necessarily normally distributed about zero with variance σ^2. Show that the least squares estimators of α and β have the smallest variance among all unbiased estimators which are linear in the y_i's. This is an application of the Markov theorem on least squares. (Use the method of Lagrangian multipliers described in Problem 11.4.)

12.3. If X and Y are random variables having variances σ_x^2 and σ_y^2 and if $E(Y \mid X = x) = \alpha + \beta x$ and $V(Y \mid X = x) = \sigma^2$, show that $\text{Cov}(X, Y) = \beta\sigma_x^2$. (Evaluate $E(XY)$ by taking the Expected value first over Y for fixed X and then over X.) Hence, by considering the identity

$$V(Y - \alpha - \beta X) = \sigma^2 = \sigma_y^2 + \beta^2\sigma_x^2 - 2\beta\,\text{Cov}(X, Y)$$

show that $\sigma^2 = \sigma_y^2(1 - \rho^2)$.

12.4. Suppose that the regression of Y on X is linear and homoscedastic as in the previous problem. Suppose furthermore that the conditional distribution of Y given $X = x$ is normal for all x, as is the marginal distribution of X. Show that the joint distribution of X and Y has the density function

$$f(x, y) = \frac{e^{-\frac{1}{2(1-\rho^2)}\left[\left(\frac{x-\mu_x}{\sigma_x}\right)^2 - 2\rho\left(\frac{x-\mu_x}{\sigma_x}\right)\left(\frac{y-\mu_y}{\sigma_y}\right) + \left(\frac{y-\mu_y}{\sigma_y}\right)^2\right]}}{2\pi\sigma_x\sigma_y\sqrt{(1-\rho^2)}}$$

This is called the *bivariate normal distribution*; the contours of constant probability density are ellipses.

12.5. Suppose that X and Y have some bivariate distribution and that the regression of Y on X, $E(Y \mid X = x) = \phi(x)$, is not necessarily linear. For purposes of prediction we may seek to find the best linear estimate of Y given X, that is to say the linear function of X, $\alpha + \beta X$, which minimises the mean square prediction error, $E(Y-\alpha-\beta X)^2$. This is called the *linear mean square regression* of Y on X. Show that it is given by

$$\beta = \frac{\text{Cov}(X, Y)}{\sigma_x^2} = \rho\frac{\sigma_y}{\sigma_x}$$

$$\alpha = \mu_y - \beta\mu_x$$

and that the minimum value of the mean square prediction error is $\sigma_y^2(1-\rho^2)$.

12.6. Suppose that k varieties of wheat have each been grown on n plots randomly chosen in a field divided into nk plots. We shall denote the yield of the ith variety on the jth plot in which it is grown by y_{ij} and we shall suppose that

$$y_{ij} = \mu_i + \varepsilon_{ij}$$

where μ_i is the true mean yield of the ith variety and ε_{ij} is an error term which is normally distributed with zero mean and variance σ^2, which is the same for all varieties. We shall define the treatment sum of squares, S_T^2, and the error sum of squares, S_E^2, by the formula

$$S_T^2 = n\sum_i (y_{i.} - y_{..})^2$$

$$S_E^2 = \sum_{i,j} (y_{ij} - y_{i.})^2$$

where $y_{i.}$ is the mean yield of the ith treatment and $y_{..}$ is the overall mean yield. Find the Expected values of S_T^2 and S_E^2 and show that S_E^2/σ^2 is a χ^2 variate with $k(n-1)$ degrees of freedom. If the μ_i's are all the same show that S_T^2/σ^2 is an independent χ^2 variate with $(k-1)$ degrees of freedom. Hence show how an F-test can be constructed to rest whether the different varieties of wheat all have the same yield.

This problem illustrates the simplest case of the Analysis of Variance known as a completely randomised design. It is in effect an extension of the t-test for testing the difference between *two* means to a situation in which we wish to test simultaneously whether there are differences between several means.

12.7. As an example of the completely randomised design, analyse the following data (extracted from Latter, 1901 and abbreviated for simplicity) on the lengths of cuckoos' eggs (in millimetres) found in nests of foster-parents of different species:

Meadow-pipit	Tree-pipit	Hedge-sparrow	Robin	Reed-warbler	Pied-wagtail	Wren
21·7	22·7	22·0	21·8	23·2	23·0	19·8
22·6	23·3	23·9	23·0	22·0	23·4	22·1
20·9	24·0	20·9	23·3	22·2	24·0	21·5
21·6	23·6	23·8	22·4	21·2	23·3	20·9
22·2	22·1	25·0	22·4	21·6	23·1	22·0
22·5	21·8	24·0	23·0	21·6	22·4	21·0
22·2	21·1	21·7	23·0	21·9	21·8	22·3
24·3	23·4	23·8	23·0	22·0	21·8	21·0
22·3	23·8	22·8	23·9	22·9	24·9	20·3
22·6	23·3	23·1	22·3	22·8	24·0	20·9

STATISTICAL TABLES

The following tables have been abridged from *Biometrika Tables for Statisticians*, Vol. 1, by permission of the *Biometrika* Trustees.

Table 1

The probability density function of the standard normal distribution

This table gives values of $\phi(z) = \frac{1}{\sqrt{2\pi}} e^{-\frac{1}{2}z^2}$; for negative values of z use the relationship $\phi(-z) = \phi(z)$.

z	$\phi(z)$	z	$\phi(z)$
0·0	·3989	2·0	·0540
0·1	·3970	2·1	·0440
0·2	·3910	2·2	·0355
0·3	·3814	2·3	·0283
0·4	·3683	2·4	·0224
0·5	·3521	2·5	·0175
0·6	·3332	2·6	·0136
0·7	·3123	2·7	·0104
0·8	·2897	2·8	·0079
0·9	·2661	2·9	·0060
1·0	·2420	3·0	·0044
1·1	·2179	3·1	·0033
1·2	·1942	3·2	·0024
1·3	·1714	3·3	·0017
1·4	·1497	3·4	·0012
1·5	·1295	3·5	·0009
1·6	·1109	3·6	·0006
1·7	·0940	3·7	·0004
1·8	·0790	3·8	·0003
1·9	·0656	3·9	·0002
2·0	·0540	4·0	·0001

TABLE 2

The cumulative probability function of the standard normal distribution

This table gives values of $\Phi(z) = \text{Prob}[Z \leq z] = \frac{1}{\sqrt{2\pi}} \int_{-\infty}^{z} e^{-\frac{1}{2}u^2} du$; for negative values of z use the relationship $\Phi(-z) = 1 - \Phi(z)$.

z	$\Phi(z)$	z	$\Phi(z)$
0·0	·5000	2·0	·9772
0·1	·5398	2·1	·9821
0·2	·5793	2·2	·9861
0·3	·6179	2·3	·9893
0·4	·6554	2·4	·9918
0·5	·6915	2·5	·9938
0·6	·7257	2·6	·9953
0·7	·7580	2·7	·9965
0·8	·7881	2·8	·9974
0·9	·8159	2·9	·9981
1·0	·8413	3·0	·9987
1·1	·8643	3·1	·9990
1·2	·8849	3·2	·9993
1·3	·9032	3·3	·9995
1·4	·9192	3·4	·9997
1·5	·9332	3·5	·9998
1·6	·9452	3·6	·9998
1·7	·9554	3·7	·9999
1·8	·9641	3·8	·9999
1·9	·9713	3·9	1·0000
2·0	·9772	4·0	1·0000

TABLE 3

Percentage points of the t distribution

This table gives values of t which are exceeded with probability P; the probability that these values will be exceeded *in either direction*, that is to say in *absolute* value, is $2P$.

	Value of P					
Degrees of freedom	·05	·025	·01	·005	·001	·0005
1	6·314	12·71	31·82	63·66	318·3	636·6
2	2·920	4·303	6·695	9·925	22·33	31·60
3	2·353	3·182	4·541	5·841	10·21	12·92
4	2·132	2·776	3·747	4·604	7·173	8·610
5	2·015	2·571	3·365	4·032	5·893	6·869
6	1·943	2·447	3·143	3·707	5·208	5·959
7	1·895	2·365	2·998	3·499	4·785	5·408
8	1·860	2·306	2·896	3·355	4·501	5·041
9	1·833	2·262	2·821	3·250	4·297	4·781
10	1·812	2·228	2·764	3·169	4·144	4·587
11	1·796	2·201	2·718	3·106	4·025	4·437
12	1·782	2·179	2·681	3·055	3·930	4·318
13	1·771	2·160	2·650	3·012	3·852	4·221
14	1·761	2·145	2·624	2·977	3·787	4·140
15	1·753	2·131	2·602	2·947	3·733	4·073
16	1·746	2·120	2·583	2·921	3·686	4·015
17	1·740	2·110	2·567	2·898	3·646	3·965
18	1·734	2·101	2·552	2·878	3·610	3·922
19	1·729	2·093	2·539	2·861	3·579	3·883
20	1·725	2·086	2·528	2·845	3·552	3·850
21	1·721	2·080	2·518	2·831	3·527	3·819
22	1·717	2·074	2·508	2·819	3·505	3·792
23	1·714	2·069	2·500	2·807	3·485	3·767
24	1·711	2·064	2·492	2·797	3·467	3·745
25	1·708	2·060	2·485	2·787	3·450	3·725
26	1·706	2·056	2·479	2·779	3·435	3·707
27	1·703	2·052	2·473	2·771	3·421	3·690
28	1·701	2·048	2·467	2·763	3·408	3·674
29	1·699	2·045	2·462	2·756	3·396	3·659
30	1·697	2·042	2·457	2·750	3·385	3·646
40	1·684	2·021	2·423	2·704	3·307	3·551
60	1·671	2·000	2·390	2·660	3·232	3·460
120	1·658	1·980	2·358	2·617	3·160	3·373
∞ (Normal)	1·645	1·960	2·326	2·576	3·090	3·291

TABLE 4

Percentage points of the χ^2 distribution

This table gives values of χ^2 which are exceeded with probability P.

Degrees of freedom	Value of P ·99	·975	·05	·025	·01	·001
1	·00016	·00098	3·84	5·02	6·63	10·83
2	·0201	·0506	5·99	7·38	9·21	13·82
3	·115	·216	7·81	9·35	11·34	16·27
4	·297	·484	9·49	11·14	13·28	18·47
5	·554	·831	11·07	12·83	15·09	20·51
6	·872	1·24	12·59	14·45	16·81	22·46
7	1·24	1·69	14·07	16·01	18·48	24·32
8	1·65	2·18	15·51	17·53	20·09	26·13
9	2·09	2·70	16·92	19·02	21·67	27·88
10	2·56	3·25	18·31	20·48	23·21	29·59
11	3·05	3·82	19·68	21·92	24·72	31·26
12	3·57	4·40	21·03	23·34	26·22	32·91
13	4·11	5·01	22·36	24·74	27·69	34·53
14	4·66	5·63	23·68	26·12	29·14	36·12
15	5·23	6·26	25·00	27·49	30·58	37·70
16	5·81	6·91	26·30	28·85	32·00	39·25
17	6·41	7·56	27·59	30·19	33·41	40·79
18	7·01	8·23	28·87	31·53	34·81	42·31
19	7·63	8·91	30·14	32·85	36·19	43·82
20	8·26	9·59	31·41	34·17	37·57	45·31
21	8·90	10·28	32·67	35·48	38·93	46·80
22	9·54	10·98	33·92	36·78	40·29	48·27
23	10·20	11·69	35·17	38·08	41·64	49·73
24	10·86	12·40	36·42	39·36	42·98	51·18
25	11·52	13·12	37·65	40·65	44·31	52·62
26	12·20	13·84	38·89	41·92	45·64	54·05
27	12·88	14·57	40·11	43·19	46·96	55·48
28	13·56	15·31	41·34	44·46	48·28	56·89
29	14·26	16·05	42·56	45·72	49·59	58·30
30	14·95	16·79	43·77	46·98	50·89	59·70

TABLE 5a

Five per cent points of the F distribution

This table gives values of F which are exceeded with a probability of ·05.

f_2 \ f_1	1	2	3	4	5	6	12	∞
1	161·4	199·5	215·7	224·6	230·2	234·0	243·9	254·3
2	18·51	19·00	19·16	19·25	19·30	19·33	19·41	19·50
3	10·13	9·55	9·28	9·12	9·01	8·94	8·74	8·53
4	7·71	6·94	6·59	6·39	6·26	6·16	5·91	5·63
5	6·61	5·79	5·41	5·19	5·05	4·95	4·68	4·36
6	5·99	5·14	4·76	4·53	4·39	4·28	4·00	3·67
7	5·59	4·74	4·35	4·12	3·97	3·87	3·57	3·23
8	5·32	4·46	4·07	3·84	3·69	3·58	3·28	2·93
9	5·12	4·26	3·86	3·63	3·48	3·37	3·07	2·71
10	4·96	4·10	3·71	3·48	3·33	3·22	2·91	2·54
11	4·84	3·98	3·59	3·36	3·20	3·09	2·79	2·40
12	4·75	3·89	3·49	3·26	3·11	3·00	2·69	2·30
13	4·67	3·81	3·41	3·18	3·03	2·92	2·60	2·21
14	4·60	3·74	3·34	3·11	2·96	2·85	2·53	2·13
15	4·54	3·68	3·29	3·06	2·90	2·79	2·48	2·07
16	4·49	3·63	3·24	3·01	2·85	2·74	2·42	2·01
17	4·45	3·59	3·20	2·96	2·81	2·70	2·38	1·96
18	4·41	3·55	3·16	2·93	2·77	2·66	2·34	1·92
19	4·38	3·52	3·13	2·90	2·74	2·63	2·31	1·88
20	4·35	3·49	3·10	2·87	2·71	2·60	2·28	1·84
21	4·32	3·47	3·07	2·84	2·68	2·57	2·25	1·81
22	4·30	3·44	3·05	2·82	2·66	2·55	2·23	1·78
23	4·28	3·42	3·03	2·80	2·64	2·53	2·20	1·76
24	4·26	3·40	3·01	2·78	2·62	2·51	2·18	1·73
25	4·24	3·39	2·99	2·76	2·60	2·49	2·16	1·71
26	4·23	3·37	2·98	2·74	2·59	2·47	2·15	1·69
27	4·21	3·35	2·96	2·73	2·57	2·46	2·13	1·67
28	4·20	3·34	2·95	2·71	2·56	2·45	2·12	1·65
29	4·18	3·33	2·93	2·70	2·55	2·43	2·10	1·64
30	4·17	3·32	2·92	2·69	2·53	2·42	2·09	1·62
40	4·08	3·23	2·84	2·61	2·45	2·34	2·00	1·51
60	4·00	3·15	2·76	2·53	2·37	2·25	1·92	1·39
120	3·92	3·07	2·68	2·45	2·29	2·17	1·83	1·25
∞	3·84	3·00	2·60	2·37	2·21	2·10	1·75	1·00

TABLE 5*b*

One per cent points of the F distribution

This table gives values of F which are exceeded with a probability of ·01

f_2 \ f_1	1	2	3	4	5	6	12	∞
1	4052	4999·5	5403	5625	5764	5859	6106	6366
2	98·50	99·00	99·17	99·25	99·30	99·33	99·42	99·50
3	34·12	30·82	29·46	28·71	28·24	27·91	27·05	26·13
4	21·20	18·00	16·69	15·98	15·52	15·21	14·37	13·46
5	16·26	13·27	12·06	11·39	10·97	10·67	9·89	9·02
6	13·75	10·92	9·78	9·15	8·75	8·47	7·72	6·88
7	12·25	9·55	8·45	7·85	7·46	7·19	6·47	5·65
8	11·26	8·65	7·59	7·01	6·63	6·37	5·67	4·86
9	10·56	8·02	6·99	6·42	6·06	5·80	5·11	4·31
10	10·04	7·56	6·55	5·99	5·64	5·39	4·71	3·91
11	9·65	7·21	6·22	5·67	5·32	5·07	4·40	3·60
12	9·33	6·93	5·95	5·41	5·06	4·82	4·16	3·36
13	9·07	6·70	5·74	5·21	4·86	4·62	3·96	3·17
14	8·86	6·51	5·56	5·04	4·69	4·46	3·80	3·00
15	8·68	6·36	5·42	4·89	4·56	4·32	3·67	2·87
16	8·53	6·23	5·29	4·77	4·44	4·20	3·55	2·75
17	8·40	6·11	5·18	4·67	4·34	4·10	3·46	2·65
18	8·29	6·01	5·09	4·58	4·25	4·01	3·37	2·57
19	8·18	5·93	5·01	4·50	4·17	3·94	3·30	2·49
20	8·10	5·85	4·94	4·43	4·10	3·87	3·23	2·42
21	8·02	5·78	4·87	4·37	4·04	3·81	3·17	2·36
22	7·95	5·72	4·82	4·31	3·99	3·76	3·12	2·31
23	7·88	5·66	4·76	4·26	3·94	3·71	3·07	2·26
24	7·82	5·61	4·72	4·22	3·90	3·67	3·03	2·21
25	7·77	5·57	4·68	4·18	3·85	3·63	2·99	2·17
26	7·72	5·53	4·64	4·14	3·82	3·59	2·96	2·13
27	7·68	5·49	4·60	4·11	3·78	3·56	2·93	2·10
28	7·64	5·45	4·57	4·07	3·75	3·53	2·90	2·06
29	7·60	5·42	4·54	4·04	3·73	3·50	2·87	2·03
30	7·56	5·39	4·51	4·02	3·70	3·47	2·84	2·01
40	7·31	5·18	4·31	3·83	3·51	3·29	2·66	1·80
60	7·08	4·98	4·13	3·65	3·34	3·12	2·50	1·60
120	6·85	4·79	3·95	3·48	3·17	2·96	2·34	1·38
∞	6·63	4·61	3·78	3·32	3·02	2·80	2·18	1·00

ANSWERS TO EXERCISES

Chapter 2

2.1. (*a*) ·160 compared with ·159; (*b*) ·175 compared with ·182.

2.2. $\frac{1}{2}$

2.3. $\frac{1}{216}$ (1, 3, 6, 10, 15, 21, 25, 27, 27, 25, 21, 15, 10, 6, 3, 1)

2.4. ·5178, ·4914

2.5. (*a*) $364 \times 363/365^2$; (*b*) $3 \times 364/365^2$; (*c*) $1/365^2$

2.6 4 cupboards are enough 95.85 per cent of the time

2.7. Yes, probability is ·000547

Chapter 3

3.4.

y \ x	−1	0	1	
0	0	$\frac{1}{3}$	0	$\frac{1}{3}$
1	$\frac{1}{3}$	0	$\frac{1}{3}$	$\frac{2}{3}$
	$\frac{1}{3}$	$\frac{1}{3}$	$\frac{1}{3}$	

If $Y = 0$, then $X = 0$ with probability 1

If $Y = 1$, then X is equally likely to be $+1$ or -1.

3.5. A cube. $\frac{1}{4}$.

Chapter 4

4.2. (*a*) mean = 15·049, median = 15·044, (*b*) mean = 19·170, median = 19·178.

4.3. $m_2 = 5{\cdot}172$

4.4. $T = n\bar{x}$, $T^2 = n^2\bar{x}^2$, $T^2/n = n\bar{x}^2$.

4.5. $m_2 = 1{\cdot}999$

4.6. mean = 15·049, $m_2 = 0{\cdot}270$

4.7. mean = 19·170, $m_2 = 0{\cdot}387$

4.8. Theoretical values, mean deviation = $\frac{1}{4}$, interquartile range = $\frac{1}{2}$, standard deviation = $1/\sqrt{12} = 0{\cdot}289$

4.9. Interquartile range = 19·598 − 18·746 = 0·852
Interquartile range/standard deviation = 1·37

4.10. Skewness = +0·17, kurtosis = 2·52

4.11. Skewness = 0, kurtosis = 2·36

4.12. (*a*) 0·249 (*b*) 0·366

Chapter 5

5.1. $\mu = 10\frac{1}{2}$, $\sigma^2 = 8\frac{3}{4}$

5.2. mean = 78.61, standard deviation = 3·16 grains

5.3. Difference: $\mu = 0$, $\sigma^2 = 6\frac{1}{4}$ sq. in.
Sum: $\mu = 11$ ft 4 in, $\sigma^2 = 18\frac{3}{4}$ sq. in.

5.4. $(X-\xi)(Y-\eta) = X(Y-\frac{2}{3})$ takes values $-\frac{1}{3}$, 0 and $\frac{1}{3}$ with equal probabilities; hence its Expected value is 0.
For lack of independence see Exercise 3.4.

5.5. $\Sigma(x_i-\bar{x})(y_i-\bar{y}) = \Sigma(x_iy_i - x_i\bar{y} - y_i\bar{x} + \bar{x}\bar{y}) = \Sigma x_iy_i - \bar{y}\Sigma x_i - \bar{x}\Sigma y_i + n\bar{x}\bar{y}$
$= \Sigma x_iy_i - n\bar{x}\bar{y} - n\bar{x}\bar{y} + n\bar{x}\bar{y} = \Sigma x_iy_i - n\bar{x}\bar{y}$.
covariance $= \cdot 122$, $r = \cdot 377$.

5.6. $\Sigma y_i = na + b\Sigma x_i$, $\bar{y} = a + b\bar{x}$, $y_i - \bar{y} = a + bx_i - (a+b\bar{x}) = b(x_i-\bar{x})$,

$$r = \frac{\Sigma(x_i-\bar{x})(y_i-\bar{y})}{\sqrt{\Sigma(x_i-\bar{x})^2\Sigma(y_i-\bar{y})^2}} = \frac{b\Sigma(x_i-\bar{x})^2}{\sqrt{b^2\Sigma(x_i-\bar{x})^2\Sigma(x_i-\bar{x})^2}} = \frac{b}{\sqrt{b^2}} = \pm 1$$

Chapter 6

6.1. ·001, ·036, ·027, ·432, ·504

6.2. ·0489, ·0751.

6.3. Observed 2·53 and 1·29, theoretical 2·50 and 1·25.

6.4. (*a*) mean $= 6 \cdot 139$, compared with 6; $p = \cdot 5116$
(*b*) $m_2 = 2 \cdot 931$, compared with (*i*) 3·000, (*ii*) 2·998,
(*c*)

No. of successes	0	1	2	3	4	5	6	7	8	9	10	11	12
Expected ($P=\frac{1}{2}$)	1	12	66	220	495	792	924	792	495	220	66	12	1
Expected ($P=p$)	1	9	55	191	450	754	921	827	541	252	79	15	1

6.5. (*a*) $2 \cdot 1 \times 10^{-6}$ compared with $2 \cdot 5 \times 10^{-6}$; (*b*) $3 \cdot 6 \times 10^{-3}$ compared with $1 \cdot 2 \times 10^{-3}$.

6.6.

No. of cells	0	1	2	3	4	5	6
Expected no. of squares	106	141	93	41	14	4	1

6.7. (*a*) ·0894, (*b*) ·0907

6.8. (*a*) $m = 16 \cdot 7$, $m_2 = 16 \cdot 4$; (*b*) $m = 26 \cdot 2$, $m_2 = 2169$. The very high variance in (*b*) is largely, but not entirely, due to a single big observation.

6.9. $m = 3 \cdot 35$ (3·26 from ungrouped data); $\sqrt{m_2} = 3 \cdot 40$

6.10. $e^{-1.5} = \cdot 223$

Chapter 7

7.1. (*a*) ·266, (*b*) ·030, (*c*) ·245

7.2. (*a*) ·721, (*b*) 40

7.4.

Head breadth (cm)	13–	13½–	14–	14½–	15–	15½–	16–	16½–
Expected frequency	2	39	304	919	1103	526	99	8

7.5. (*a*) ·842, (*b*) ·976 compared with theoretical values of ·841 and ·977

7.6. ·977

7.7. If x_i is no. of cupboards required by *i*th chemist, then $x = \Sigma x_i$, is approximately normal with $\mu = 50E(x_i) = 25$, $\sigma^2 = 50V(x_i) = 22 \cdot 5$. $\mu + 1 \cdot 645\sigma = 33$.

7.8. ·179; about 3452
7.9. About 170.
7.10. About 400.

Chapter 8

8.1. About ·025 since Prob $[\chi^2_{[9]} \geqslant 2{\cdot}70] = {\cdot}975$
8.2. About ·002 since the chance that a t variate with 3 d.f. will exceed 10·21 is ·001.
8.3. About ·05 since 4·28 is the upper 5 per cent point of an F variate with 6 and 6 d.f.

Chapter 9

9.1. $d = 0{\cdot}63$, $P = 0{\cdot}53$ (two-tailed). Not significant.
9.2. $d = 2{\cdot}83$, $P = 0{\cdot}0023$ (one-tailed). Significant at 1 per cent level.
9.3. $d = 1{\cdot}21$, $P = 0{\cdot}22$ (two-tailed). Not significant
This can also be treated as a 2×2 table, $\chi^2_{[1]} = 1{\cdot}47$.
9.4. $t = 1{\cdot}69$ with 10 d.f., $P > 0{\cdot}10$ (two-tailed, since it is conceivable that kiln-drying will reduce yield)
9.5. (*a*) $t = 2{\cdot}06$ with 4 d.f., $\cdot 05 < P < \cdot 10$ (one-tailed)
(*b*) $t = 13{\cdot}75$ with 4 d.f., highly significant
(*c*) $t = 1{\cdot}44$ with 8 d.f., not significant
9.6. (*a*) $d = 5{\cdot}14$, highly significant
(*b*) $\chi^2 = 7{\cdot}0$ with 9 d.f. (combining $0-1$ and $11-12$); not significant
9.7. (*a*) $\chi^2 = 2{\cdot}4$ with 2 d.f. (combining 3+); not significant
(*b*) $\chi^2 = 61{\cdot}9$ with 2 d.f. (combining 3+); highly significant
(*c*) $\chi^2 = 3{\cdot}2$ with 4 d.f. (combining 5+); not significant
9.8. (*a*) $\chi^2 = 9{\cdot}9$ with 13 d.f. (combining $\leqslant 58''$ & $73''+$); not significant
(*b*) $\chi^2 = 8{\cdot}24$ with 4 d.f. (combining $13-$ with $13\frac{1}{2}-$); not significant at 5 per cent level.
9.9. (*a*) $\chi^2 = 94$ with 5 d.f.; highly significant
(*b*) $\chi^2 = 271$ with 5 d.f.; highly significant
(*c*) $\chi^2 = 18{\cdot}8$ with 25 d.f.; not significant
9.10. $$a - \frac{(a+b)(a+c)}{n} = \frac{na - a(a+b+c) - bc}{n} = \frac{na - a(n-d) - bc}{n} = \frac{ad-bc}{n}$$

and likewise for other values of $(0-E)$. Hence

$$\chi^2 = \sum \frac{(0-E)^2}{E}$$

$$= \frac{(ad-bc)^2}{n} \left\{ \frac{(c+d)(b+d) + (a+c)(c+d) + (a+b)(b+d) + (a+b)(a+c)}{(a+b)(c+d)(a+c)(b+d)} \right\}$$

$$= \frac{n(ad-bc)^2}{(a+b)(c+d)(a+c)(b+d)}$$

9.11. $\chi^2 = 13{\cdot}0$ with 1 d.f., highly significant

9.12. From Exercise 10, $\left| a - \frac{(a+b)(a+c)}{n} \right| - \frac{1}{2} = \frac{| ad - bc |}{n} - \frac{1}{2}$, from which formula follows as in that Exercise.

$\chi^2 = 10{\cdot}5$ with 1 d.f., highly significant.

Chapter 10

10.1. $\cdot 1966 \pm \cdot 0055$
10.2. (*a*) $\cdot 0234 \pm \cdot 0005$, (*b*) $\cdot 0224 \pm \cdot 0005$, (*c*) $\cdot 0010 \pm \cdot 0007$
10.3. $6{\cdot}125 \pm \cdot 156$
10.4. 208 ± 40 million
10.5. (*a*) $\cdot 75 \pm 1{\cdot}28$, (*b*) $2{\cdot}33 \pm 1{\cdot}43$, (*c*) $1{\cdot}58 \pm \cdot 88$
10.6. $1{\cdot}38 \leqslant \sigma \leqslant 3{\cdot}65$
10.7. (*a*) $15{\cdot}40 \pm 3{\cdot}11$, (*b*) $2{\cdot}00 \pm 3{\cdot}21$
10.8. $\cdot 899$

Chapter 12

12.1. $\Sigma x_i = 25$, $\Sigma y_i = 262$, $\Sigma x_i^2 = 75$, $\Sigma x_i y_i = 454$, $\Sigma y_i^2 = 3604$; hence $\Sigma(x_i - \bar{x})^2 = \Sigma x_i^2 - n\bar{x}^2 = 50$, $\Sigma(x_i - \bar{x})(y_i - \bar{y}) = \Sigma x_i y_i - n\bar{x}\bar{y} = 192$, $b = 192/50 = 3{\cdot}84$, $a = \bar{y} - b\bar{x} = 10{\cdot}48 - 3{\cdot}84 = 6{\cdot}64$.
Best fitting straight line is: $y = 6{\cdot}64 + 3{\cdot}84x$.
12.2. $S^2 = 120{\cdot}96$
12.3. $\beta = 3{\cdot}84 \pm \cdot 67$, $\alpha = 6{\cdot}64 \pm 1{\cdot}16$
12.4. $y = 9{\cdot}010 + \cdot 315x$, $x = 12{\cdot}368 + \cdot 452y$
12.5. $\cdot 345 \leqslant \rho \leqslant \cdot 407$
12.6. $r = +\cdot 61$, $t = r\sqrt{(n-2)}/\sqrt{(1-r^2)} = 2{\cdot}17$ with 8 degrees of freedom, which is not significant at the 5 per cent level if a two-tailed test is used. A positive correlation is expected since (1) the responses to the two drugs are measured from the *same* control period of sleep, (2) some patients are likely to be more responsive to both drugs than others.

SOLUTIONS TO PROBLEMS

Chapter 2

2.1. Put $A = E_{n-1}$ or E_n, write down addition law for $E_1, \ldots, E_{n-2}, A$ and then substitute back for A.
If in doubt, try $n = 3$ first.

2.2. Use generalised Venn diagram (see Fig. 2).
$\cdot 0006433 < P < \cdot 0006435$.

2.4. Brothers ·59, unrelated men ·38

2.5.

	Purple-flowered	Red-flowered
Long pollen	$\frac{1}{4}(2+\theta)$	$\frac{1}{4}(1-\theta)$
Round pollen	$\frac{1}{4}(1-\theta)$	$\frac{1}{4}\theta$

2.6. (*a*) 15/128, (*b*) 3/16

2.7. ·096, ·497, ·407

Chapter 3

3.1. (*a*) $g(y) = \frac{1}{2}y^{-\frac{1}{2}}$, (*b*) $g(y) = 2y$, $0 \leqslant y \leqslant 1$.

3.2. (*a*) $g(y) = -\dfrac{1}{b}f\left(\dfrac{y-a}{b}\right)$, (*b*) $g(y) = e^{-y}$, $0 \leqslant y < \infty$

3.3. $g(y) = \frac{1}{3}y^{-\frac{1}{2}}$ for $0 \leqslant y < 1$, $g(y) = \frac{1}{6}y^{-\frac{1}{2}}$ for $1 < y \leqslant 4$.

3.5. $h(u) = u$, $0 \leqslant u \leqslant 1$, $h(u) = 2-u$, $1 \leqslant u \leqslant 2$

3.6. $h(u) = \frac{1}{2}$, $0 \leqslant u \leqslant 1$, $h(u) = \frac{1}{2}u^{-2}$, $1 \leqslant u < \infty$

3.7. If ship lies in cocked hat when errors are θ_1, θ_2 and θ_3, it will also lie in cocked hat when errors are $-\theta_1$, $-\theta_2$ and $-\theta_3$ but not when only some of the signs are changed; hence ship will lie in 2 of the 8 equally likely cocked hats formed from $\pm\theta_1$, $\pm\theta_2$, $\pm\theta_3$.

3.8. $255\frac{1}{2}$ miles

Chapter 4

4.1. Equality holds if $g(x)/h(x) = $ constant.

4.2. (*a*) $g(x) = \sqrt{f(x)}$, $h(x) = (x-\mu)^2\sqrt{f(x)}$,
(*b*) $g(x) = (x-\mu)\sqrt{f(x)}$, $h(x) = (x-\mu)^2\sqrt{f(x)}$

4.3. mean $= \frac{1}{2}(\mu_1+\mu_2)$, variance $= \sigma^2+\frac{1}{4}(\mu_1-\mu_2)^2$, skewness $= 0$,

kurtosis $= 3 - \dfrac{2(\mu_1-\mu_2)^2}{[4\sigma^2+(\mu_1-\mu_2)^2]^2}$, bimodal if $|\mu_1-\mu_2| > 2\sigma$.

4.4. (*a*) If x_1, x_2, x_3 are three observations in ascending order and if c lies between x_1 and x_3, then $|x_1-c|+|x_2-c|+|x_3-c| = x_3-x_1+|x_2-c|$, which is a minimum when $c = x_2$. This argument can be extended to any number of observations; for an even number there is a minimum for any value of c between the two middle observations.
(*b*) $\Sigma(x_i-c)^2 = \Sigma[(x_i-\bar{x})-(c-x)]^2 = \Sigma(x_i-\bar{x})+n(c-\bar{x})^2$
which is a minimum when $c = \bar{x}$.

4.5. Write $g(x) = (x-\mu)^2$, $a = k^2\sigma^2$.

CHAPTER 5

5.1. $\mu_3(X+Y)=\mu_3(X)+\mu_3(Y)$
$\mu_4(X+Y)=\mu_4(X)+\mu_4(Y)+6\mu_2(X)\,.\,\mu_2(Y)$

5.2. $\rho=h/k$. $E(Y-X)^2=2V(Y)(1-\rho)$

5.5. Estimate of $g=31{\cdot}99$ ft/sec²; standard error $=\cdot036$

5.8. If c stands for coefficient of variation,

$$c(u)=\tfrac{1}{2}\sqrt{c^2(p)+c^2(s_1)+c^2(d)+c^2(s_2)}$$

CHAPTER 6

6.1. $G(s)=(Q+Ps)^n$, $\mu_{[r]}=n(n-1)\ldots(n-r+1)P^r$

6.2. $G(s)=e^{\mu(s-1)}$, $\mu_{[r]}=\mu^r$.

6.5. If number of nests per unit area is d, then the probability that there will be no nests within a distance x of some fixed point is $e^{-\lambda\pi x^2}$ since the area of the circle with radius x is πx^2, and the probability that there will be a nest between x and $x+dx$ units is $2\pi\lambda x dx$, since the area of the annulus is $2\pi x dx$. Hence the density function of X, the distance of the nearest nest, is $e^{-\lambda\pi x^2}\,2\pi\lambda x$, and the density function of $Y=X^2$ is $\lambda\pi e^{-\lambda\pi y}$.

6.7. Find the number of ways of drawing x balls out of R red balls and of simultaneously drawing $(n-x)$ balls out of $(N-R)$ black balls, and divide it by the total number of ways of drawing n balls out of N balls.

	0	1	2	3
$P(x)$ (binomial)	·216	·432	·288	·064
$P(x)$ (hypergeometric)	·167	·500	·300	·033

6.8. $E(Z_i)=P$, $V(Z_i)=PQ$, Cov $(Z_i, Z_j)=-PQ/(N-1)$.
In Problem 6.7, mean $=1{\cdot}2$, variance $=\cdot72$ (with replacement), $=\cdot56$ (without replacement).

6.9. $G(s)=P/(1-Qs)$, $\mu_{[r]}=r!(Q/P)^r$, $\mu=Q/P$, $\mu_2=Q/P^2$, skewness $=(Q+1)/\sqrt{Q}$.

6.10. Cov $(X_i, X_j)=-nP_iP_j$.

CHAPTER 7

7.3. If $X=\Sigma X_i$, $E(X_i)=0$, $E(X_i^r)=\mu_r$, show, by expanding $(X_1+X_2\ldots+X_n)^r$ and taking Expected values, that $E(X^3)=n\mu_3$, $E(X^4)=n\mu_4+3n(n-1)\mu_2^2$.

7.6. Error $=P[Y\geqslant rX\ \&\ X<0]-P[Y\leqslant rX\ \&\ X<0]$ which is less in absolute magnitude than the sum of these probabilities, which is $P[X<0]$.

CHAPTER 8

8.4. $V(S^2)=\dfrac{\mu_4}{n}(n-1)^2-\dfrac{\mu_2^2}{n}(n-1)(n-3)$.

8.5. $\text{Cov}\,(S^2, \bar{x}) = \dfrac{(n-1)}{n}\mu_3$

8.6. $g(v) = 2f^{\frac{1}{2}f}v^{f-1}e^{-\frac{1}{2}fv^2}/A(f) \qquad 0 \leqslant v < \infty$
For the density function of T, see p. 133

8.7. See p. 135

8.8. mean $= \dfrac{p}{p+q}$, variance $= \dfrac{pq}{(p+q)^2(p+q+1)}$

8.9. $G(s) = P^n/(1-Qs)^n$, mean $= nQ/P$, variance $= nQ/P^2$.
[Note that the p.g.f. is the nth power of the p.g.f. of the geometric distribution found in Problem 6.9; it follows that, if n is an integer, the negative binomial distribution can be interpreted as the distribution of the sum of n independent geometric variables, or as the waiting time until the nth success.]
By equating the observed with the theoretical mean and variance in Table 15 on p. 96, we find $P = {\cdot}68$, $Q = {\cdot}32$, $n = 1$, so that the negative binomial is in this case the geometric distribution.

No. of accidents	0	1	2	3	4	5
Expected no. of women	440	141	45	14	5	1

Chapter 9

9.1. $n = 9$, $c = 2{\cdot}368$

9.3. (a) $P = {\cdot}00047$, (b) $P = {\cdot}00054$. The one-tailed test is appropriate since we can predict beforehand that the degree of concordance will be at least as high in monozygotic as in dizygotic twins; note the asymmetry of the tails.

9.4. $\chi^2_{[6]} = 11{\cdot}4, \quad {\cdot}05 < P < {\cdot}10$

Chapter 10

10.1. $(a-\rho b)^2 \leqslant t^{*2}s^2(c_1+\rho^2c_2-2\rho c_3)$, where t^* is the appropriate percentage point of the t distribution with f degrees of freedom.
$\rho = 3{\cdot}11 \pm 1{\cdot}02$

Chapter 11

11.1. median follows Beta distribution with $p = q = \frac{1}{2}(n+1)$, mean $= \frac{1}{2}$,
variance $= \dfrac{1}{4(n+2)}$ compared with approximate value of $\dfrac{1}{4n}$.

11.2.

Estimator	Mean	Median	$(n+1)y/(2n+1)$
Variance	$\dfrac{\mu^2}{3(2n+1)}$	$\dfrac{\mu^2}{(2n+3)}$	$\dfrac{\mu^2}{(2n+1)(2n+3)}$

11.3. $\hat{\mu} = -\dfrac{1}{x}\log\dfrac{a}{n}$, $V(\hat{\mu}) = (e^{\mu x}-1)/nx^2$, which is a minimum when $\mu x = 1{\cdot}6$; x should be chosen so that there are between 1 and 2 organisms per sample.

11.4. $w_i = \frac{1}{\sigma_i^2} \Big/ \Sigma(1/\sigma_i^2)$, minimum variance $= 1/\Sigma(1/\sigma_i^2)$.

$$\Sigma w_i t_i = \frac{\Sigma n_i \bar{x}_i}{\Sigma n_i} = \frac{\text{sum of all observations}}{\text{Total number of observations}}$$

11.5. $V\left(\frac{m}{nM}\right) = \frac{1}{nN}\left[\frac{1}{M} - \frac{1}{N}\right]\left[1 - \frac{n-1}{N-1}\right]$, which can be calculated if we assume that $m/nM = \frac{1}{N}$.

95 per cent confidence limits for N are 641 to 2,268.

11.6. $P(n) = \frac{M}{N}\binom{M-1}{m-1}\binom{N-M}{n-m}\Big/\binom{N-1}{n-1} \qquad m \leqslant n \leqslant N+m-M$

$$E(n) = \frac{m(N+1)}{M+1}, \quad E[n(n+1)] = \frac{m(m+1)(N+1)(N+2)}{(M+1)(M+2)}$$

(To perform the summation note that $\Sigma P(n) = 1$.)

$\frac{n(M+1)}{m} - 1$ is an unbiased estimator of N with variance

$$\frac{(M-m+1)(N+1)(N-M)}{m(M+2)}$$

11.7. $\hat{\theta} = \cdot 784$, 95 per cent confidence limits $\cdot 758 - \cdot 810$, hence 95 per cent confidence limits for π are $\cdot 100 - \cdot 129$.

Chapter 12

12.1. $V(y_{n+1} - y^*_{n+1}) = \sigma^2\left[1 + \frac{1}{n} + \frac{(x_{n+1}-\bar{x})^2}{\Sigma(x_i-\bar{x})^2}\right]$;

estimate σ^2 by residual mean square with $(n-2)$ degrees of freedom, and hence place a confidence interval on y_{n+1}.

12.6. $E(S_T^2) = n\sum_i (\mu_i - \mu)^2 + (k-1)\sigma^2$, where $\mu = \Sigma\mu_i/k$

$E(S_E^2) = k(n-1)\sigma^2$

If μ_i's are equal, the ratio

$$F = \frac{S_T^2/(k-1)}{S_E^2/k(n-1)}$$

should follow the F distribution with $(k-1)$ and $k(n-1)$ degrees of freedom.

12.7. $S_T^2 = 29.9$

$S_E^2 = 50.9$

$F = 6.16$ with 6 and 63 degrees of freedom, highly significant.

REFERENCES

BAILEY, N. T. J. 1951. *Biometrika*, **38**, 293.

1959. *Statistical Methods in Biology*. E.U.P., London.

1964. *The Elements of Stochastic Processes with applications to the Natural Sciences*. Wiley, New York.

BARTLETT, M. S. 1960. *Stochastic Population Models*. Methuen, London.

BATESON, W. 1909. *Mendel's Principles of Heredity*. University Press, Cambridge. (Contains a translation of Mendel's papers.)

BAYES, T. 1763. *Phil. Trans. Roy. Soc.*, **53**, 370. (Reproduced 1958, in *Biometrika*, **45**, 293.)

BERKSON, J. 1938. *J. Amer. Statist. Ass.*, **33**, 526.

BERKSON, J., MAGATH, T. B. and HURN, M. 1935. *J. Amer. Statist. Ass.*, **30**, 414.

BERNOULLI, J. 1713. *Ars conjectandi, opus posthumum*. Basle.

BLACKWELL, D. and GIRSHICK, M. A. 1954. *Theory of Games and Statistical Decisions*. Wiley, New York.

BORTKIEWICZ, L. VON. 1898. *Das Gesetz der kleinen Zahlen*. Leipzig.

BOWEN, M. G. 1947. *J. Agric. Res.*, **75**, 259.

BRAITHWAITE, R. B. 1955. *Scientific Explanation*. University Press Cambridge.

BROWN, B. H. 1919. *Amer. Math. Monthly*, **26**, 351.

BUTLER, V. J. 1961. In *Anglo-Saxon Coins*, pp. 195-214, ed. R. H. M. Dolley. Methuen, London.

CARNAP, R. 1950. *Logical Foundations of Probability*. Routledge and Kegan Paul, London.

CHAMBERLAIN, A. C. and TURNER, F. M. 1952. *Biometrics*, **8**, 55.

CHERNOFF, H. and MOSES, L. E. 1959. *Elementary Decision Theory*. Wiley, New York.

COTTON, C. 1674. *The Compleat Gamester*.

COURANT, R. 1934. *Differential and Integral Calculus*. Blackie, London & Glasgow.

CZUBER, E. 1903. *Wahrscheinlichkeitsrechnung und ihre Anwendung auf Fehlerausgleichung, Statistik und Lebensversicherung*, Leipzig.

DE MOIVRE, A. 1733. *Approximatio ad Summam Terminorum Binomii* $(a+b)^n$ *in Seriem expansi*. A translation by de Moivre is reproduced in D. E. Smith's *Source Book in Mathematics* (McGraw-Hill, 1929).

FATT, P. and KATZ, B. 1952. *J. Physiol.*, **117**, 109.

FERRAR, W. L. 1941. *Algebra*. University Press, Oxford.

FISHER, R. A. 1915. *Biometrika*, **10**, 507.
1921. *Metron*, **1**, Part 4, p. 3.
1924. *Proc. Int. Math. Congress*, Toronto, p. 805.
1925. *Statistical Methods for Research Workers.* Oliver & Boyd, Edinburgh.
1925. *Metron*, **5**, 90.
1935. *Ann. Eugen.*, **6**, 391.
1956. *Statistical Methods and Scientific Inference.* Oliver & Boyd, Edinburgh.

FISHER, R. A. and YATES, F. 1953. *Statistical Tables for Biological, Agricultural and Medical Research.* (4th edn.) Oliver & Boyd, Edinburgh.

GALTON, F. 1889. *Natural Inheritance.* (1st edn.) Macmillan, London.

GAUSS, C. F. 1809. *Theoria Motus Corporum Coelestium.* Hamburg.

GEISSLER, A. 1889. *Z.K. sächsischen statistischen Bureaus*, **35**, 1.

GREENWOOD, A. W., BLYTH, J. S. S. and CALLOW, R. K. 1935. *Biochem. J.*, **29**, 1400.

GREENWOOD, M. and YULE, G. U. 1920. *Jour. Roy. Stat. Soc.*, **83**, 255.

HARRIS, H. and KALMUS, H. 1949. *Ann. Eugen.*, **15**, 24.

HARTMANN, G. 1939. *Ann. Eugen.*, **9**, 123.

HELMERT, F. R. 1876. *Zeit. für Math. und Phys.*, **21**, 192.
1876. *Astronomische Nachrichten*, **88**, No. 2096.

HUFF, D. 1960. *How to Take a Chance.* Gollancz, London.

JEFFREYS, H. 1961. *Theory of Probability.* (3rd edn.) University Press, Oxford.

KERRICH, J. E. 1946. *An Experimental Introduction to the Theory of Probability.* Munksgaard, Copenhagen.

KEYNES, J. M. 1921. *A Treatise on Probability.* Macmillan, London.

KING, H. D. 1924. *Anat. Record*, **27**, 337.

LAPLACE. 1812. *Théorie analytique des Probabilités*, Paris.
1814. *Essai Philosophique sur les Probabilités.* (English translation by Truscott & Emory (1902), reprinted 1951 by Dover Publications.)

LATTER, O. H. 1901. *Biometrika*, **1**, 69.

LURIA S. E. AND DELBRÜCK, M. 1943. *Genetics*, **28**, 491.

MACDONELL, W. R. 1901. *Biometrika*, **1**, 177.

MCKINSEY, J. 1952. *Introduction to the Theory of Games.* McGraw-Hill, New York.

MARSDEN, E. and BARRATT, T. 1911. *Proc. Phys. Soc.*, **23**, 367.

MENDEL, G. See BATESON, 1909.

MISES, R. VON. 1957. *Probability, Statistics and Truth.* Allen and Unwin, London.

NAIR, A. N. K. 1941. *Sankhyā*, **5**, 383.

PEARSON, K. 1900. *Phil. Mag.*, **50**, 157.

PEARSON, K. and LEE, A. 1903. *Biometrika*, **2**, 357.

POINCAIRÉ, H. 1896. *Calcul des Probabilités.* Gauthier-Villars, Paris.

POISSON, S. D. 1837. *Recherches sur la probabilité des Jugements en Matière Criminelle et en Matière Civile.* Paris.

QUETELET, L. A. J. 1835. *Essai de Physique Sociale.* Bruxelles.

RAMSEY, F. P. 1931. *The Foundations of Mathematics and other Logical Essays.* Routledge and Kegan Paul, London and New York.

REICHENBACH, H. 1949. *The Theory of Probability.* University of California Press, Los Angeles.

RUTHERFORD, E., GEIGER, H. and BATEMAN, H. 1910. *Phil. Mag.*, **20**, 698.

SAVAGE, L. J. 1954. *The Foundations of Statistics.* Wiley, New York.

SIEGEL, S. 1956. *Non-parametric Statistics.* McGraw-Hill, New York.

SINNOTT, E. W. 1937. *Proc. Nat. Acad. Sci. Wash.*, **23**, 224.

SMITH, D. A. 1929. *A Source Book in Mathematics.* McGraw-Hill, New York.

'STUDENT'. 1907. *Biometrika,* **5**, 351.

'STUDENT'. 1908. *Biometrika,* **6**, 1.

THOMPSON, W. D'ARCY. 1917 (1st edn.) (2nd edn. 1942. reprinted 1952) *On Growth and Form.* University Press, Cambridge.

WALD, A. 1939. *Ann. Math. Stat.*, **10**, 299.
1950. *Statistical Decision Functions.* Wiley, New York.

WEINER, J. S. 1955. *The Piltdown Forgery.* University Press, Oxford.

WILLIAMS, E. J. 1959. *Regression Analysis.* Wiley, New York.

WILLIAMS, J. D. 1954. *The Compleat Strategyst.* McGraw-Hill, New York.

YULE, G. U. and KENDALL, M. G. 1950. *An Introduction to the Theory of Statistics.* (14th edn.) Griffin, London.

INDEX